Supporting Combat Power Projection Away from Fixed Infrastructure

JAMES A. LEFTWICH, BRADLEY DEBLOIS, DAVID T. ORLETSKY

Prepared for the Department of the Air Force
Approved for public release; distribution unlimited

PROJECT AIR FORCE

For more information on this publication, visit **www.rand.org/t/RRA596-1**.

About RAND

The RAND Corporation is a research organization that develops solutions to public policy challenges to help make communities throughout the world safer and more secure, healthier and more prosperous. RAND is nonprofit, nonpartisan, and committed to the public interest. To learn more about RAND, visit www.rand.org.

Research Integrity

Our mission to help improve policy and decisionmaking through research and analysis is enabled through our core values of quality and objectivity and our unwavering commitment to the highest level of integrity and ethical behavior. To help ensure our research and analysis are rigorous, objective, and nonpartisan, we subject our research publications to a robust and exacting quality-assurance process; avoid both the appearance and reality of financial and other conflicts of interest through staff training, project screening, and a policy of mandatory disclosure; and pursue transparency in our research engagements through our commitment to the open publication of our research findings and recommendations, disclosure of the source of funding of published research, and policies to ensure intellectual independence. For more information, visit www.rand.org/about/principles.

RAND's publications do not necessarily reflect the opinions of its research clients and sponsors.

Published by the RAND Corporation, Santa Monica, Calif.

Library of Congress Cataloging-in-Publication Data is available for this publication.
ISBN: 978-1-9774-0801-3

Cover: Rick Penn-Kraus; photo: U.S. Air Force/Staff Sgt. Joshua King.

Preface

Recognizing the challenge of competing against near-peer adversaries with significant inventories of missiles, the Air Force Warfighting Integration Capability (AFWIC) has been exploring alternative future force designs intended to better position the Joint Force to deter and defeat aggression in highly contested environments. One option that AFWIC is considering is the use of a low-cost attritable[1] aircraft technology (LCAAT) class of weapon system. Designed to operate as a stand-in force, the LCAAT class of weapon system is an unmanned aircraft that can carry weapons and will allow the U.S. Air Force to project combat power with limited reliance on vulnerable runways.

In 2019, RAND Project AIR FORCE conducted preliminary research and analysis on the mission generation and support requirements associated with the LCAAT class of weapon system.[2] Building on that work, this report focuses on the extended logistics and sustainment network required for LCAATs operating from forward locations away from runways and fixed infrastructure; explores the manpower and logistics support footprint associated with different mission generation and support concepts; and highlights areas for future analysis,

---

[1] When we were conducting our analysis, the Air Force was referring to this class of weapon system as the low-cost attritable aircraft technology. More recently, the Air Force has started to refer to this class as affordable runway-independent unmanned aerial vehicles. For this report, we retain the use of LCAAT.

[2] See James A. Leftwich, *Low-Cost Attritable Aircraft Technology: Logistics Concept of Support for Deployment and Employment*, Santa Monica, Calif.: RAND Corporation, 2020. Not available to the general public.

experimentation, and engineering design modifications. The research reported here was conducted within the Strategy and Doctrine Program of RAND Project AIR FORCE as part of a fiscal year 2020 project.

The audience for this report is primarily AFWIC leadership and team members, but it should also be of interest to Air Force personnel developing the Agile Combat Employment concept, as well as other technologists, planners, and operators within the Department of Defense who are focused on developing and fielding improved capabilities for challenging warfighting scenarios.

RAND Project AIR FORCE

RAND Project AIR FORCE (PAF), a division of the RAND Corporation, is the Department of the Air Force's (DAF's) federally funded research and development center for studies and analyses, supporting both the United States Air Force and the United States Space Force. PAF provides the DAF with independent analyses of policy alternatives affecting the development, employment, combat readiness, and support of current and future air, space, and cyber forces. Research is conducted in four programs: Strategy and Doctrine; Force Modernization and Employment; Resource Management; and Workforce, Development, and Health. The research reported here was prepared under contract FA7014-16-D-1000.

Additional information about PAF is available on our website: www.rand.org/paf/

This report documents work originally shared with the DAF on September 30, 2020. The draft report, issued on October 27, 2020, was reviewed by formal peer reviewers and DAF subject-matter experts.

Contents

Figures and Tables

Figures

Tables

Summary

Faced with the challenge of deterring and defeating aggression by the sorts of highly capable adversaries highlighted in the 2018 National Defense Strategy, the U.S. Air Force (USAF) is exploring alternative weapon systems and concepts of employment that will allow it to generate combat power without being harnessed to air bases and runways that the adversary likely will view as high-value targets. Termed low-cost attritable aircraft technology (LCAAT),[3] this unmanned class of system can carry munitions and is intended to operate in remote and austere locations and be launched using a variety of means that do not include long takeoff runs. Operating with runway independence presents a unique set of challenges not only for aerospace engineers and technologists but also for the logisticians responsible for supporting and sustaining these types of expeditionary operations.

In exploring alternatives for different support functions, we used a "capability–concept of operations" structure. We started with current capabilities in the USAF inventory and current concepts of operations. We then looked at other, nonmainstream USAF capabilities, as well as current capabilities that exist in other military services and commercially, and explored their use in new concepts of operation that are not common in the USAF. Finally, we considered new capabilities

[3] When we were conducting our analysis, the Air Force was referring to this class of weapon system as the low-cost attritable aircraft technology. More recently, the Air Force has started to refer to this class as affordable runway-independent unmanned aerial vehicles. For this report, we retain the use of LCAAT.

that could be achieved through engineering design modifications to the LCAAT and employed those using new concepts of operation.

Because the USAF has grown accustomed to operating from main operating bases in sanctuary positions, few capabilities in its inventory of combat support and combat service support equipment are suited to remote operations. Capabilities required to operate in austere environments do exist in pockets of the USAF (e.g., special forces and contingency response groups), but they are not mainstream, and their tactics, techniques, and procedures (TTPs) are not broadly embraced by conventional Air Force units. Logisticians supporting LCAAT operations can leverage the unique resources and TTPs of these specialized units, as well as those of other services like the U.S. Marine Corps, to support expeditionary combat operations with a light footprint in a manner that improves the survivability of the combat and support forces.

Key findings of this analysis are the following:

- The time required to recover an LCAAT and prepare it for its next mission determines the amount of combat power that an LCAAT-equipped unit can deliver and is the largest determinant of personnel and equipment requirements.
  - Personnel and equipment requirements for LCAAT operations are considerably less than those for traditional platforms.
  - For a similar level of weapon delivery, the LCAAT operation requires approximately 20–60 percent of the personnel and 40–65 percent of the equipment required for a traditional F-16 operation, depending on LCAAT turn time.
- Sustainment requirements for LCAAT operations are significant. LCAAT takeoff and recovery methods currently being considered by the USAF (i.e., rocket-assisted takeoff [RATO] rockets and parachute/airbags) contribute significantly to the sustainment footprint. The tradeoff for that footprint is increased resiliency through dispersal and runway independence.
  - Eliminating the need for RATO rockets could reduce the daily sustainment requirements by 50 percent.
- Few traditional USAF capabilities (e.g., basic expeditionary airfield resources [BEAR], R-11 refuelers, fuels operational readi-

ness capability equipment [FORCE]) are suited for the type of expeditionary, runway-independent operations envisioned for the LCAAT. For a given turn time, alternative, nontraditional support concepts can reduce the total footprint and increase resilience by enabling distributed operations.
- In considering future force designs, logistics and sustainment analysis early in the process provides benefits to both the research and acquisition communities; the analysis can be used to inform engineering design modifications and future equipment requirements that better enable deployment and employment.

In light of those findings, the following are offered to the Air Force for consideration:

- Continue to pursue a version of an LCAAT as part of a future force design that is capable of operating decoupled from fixed infrastructure.
- Continue to explore launch methods that are mobile, do not require a takeoff run, and eliminate the need for RATOs.
- Pursue recovery methods that are precision guided and can reduce the time required to prepare the air vehicle for its next mission.
- Aggressively pursue (perhaps through extensive field tests and experiments) ways to shorten the high turn-time drivers.
- Institutionalize the use of nonmainstream capabilities like the helicopter expedient refueling systems and airfield rapid response kit.
- Actively engage the combat support research (for engineering design considerations) and acquisition (for acquiring uniquely capable equipment) communities as a part of future force design.
- Institutionalize logistics and sustainment analysis as a part of the force design process.

Acknowledgments

We thank Lt Gen Timothy Fay for sponsoring this work when he was AF/A5. The authors are thankful to members of the Air Force Warfighting Integration Capability's ARIUAVs logistics modeling team for generously setting aside time to meet virtually for extended conversations and information sharing. These include Maj Kevin Smyth, SMSgt Jason Kraemer, and Eloy Rodriguez. We also thank Bill Baron, Doug Meador, Vince Raska, and John Sletton for providing feedback and sharing their work at the Air Force Research Laboratory on the XQ-58A program. Additionally, we thank Maj Justin Bateman from the 6th Security Forces Squadron and CMSgt Scott Grabham from the 53rd Weapons Evaluation Group for providing input on combat service support requirements. Finally, we thank our RAND colleagues David Ochmanek and John Drew for feedback on the analysis throughout the project, and Jeff Hagen and Kristin Lynch for their thoughtful reviews.

Acknowledgment of these individuals does not imply their endorsement of the views expressed in this report.

Abbreviations

AFRL	Air Force Research Laboratory
AFWIC	Air Force Warfighting Integration Capability
ARRK	air rapid response kit
BEAR	basic expeditionary airfield resources
BOS	base operating support
C2	command and control
CONOP	concept of operations
CS	combat support
CSS	combat service support
EMALS	electromagnetic aircraft launch system
EOD	explosive ordnance disposal
FOL	forward operating location
FORCE	fuels operational readiness capability equipment
HERS	helicopter expedient refueling system
KE	kinetic energy
kW	kilowatt
LCAAT	low-cost attritable aircraft technology
MJ	megajoule
NSM	naval strike missile
PAF	Project AIR FORCE
QUP	quantity unit pack

RADR	rapid airfield damage repair
RATO	rocket-assisted takeoff
SDB	small-diameter bomb
SME	subject-matter expert
SOF	special operation forces
STON	short ton
UAS	unmanned aerial system
UAV	unmanned aerial vehicle
USAF	U.S. Air Force
USN	U.S. Navy
UTC	unit type code

CHAPTER ONE
Introduction

The 2018 National Defense Strategy highlights that the United States is facing adversaries that are increasingly becoming capable of denying combat forces access to the battlespace and inflicting significant damage on fixed infrastructure from which U.S. forces operate. Years of RAND Project AIR FORCE (PAF) analysis have modeled the extent to which adversary attacks on base infrastructure can significantly diminish the ability of the Air Force to generate combat power. The research highlights that attacks on runways, fuel and munition storage facilities, parked aircraft, and lodging facilities can severely degrade combat power projection.[1]

LCAAT Concept of Employment Overview

Since 2017, the Air Force Warfighting Integration Capability (AFWIC) has explored concepts for a future force design that would better enable the Air Force to compete against near-peer adversaries capable of denying access to the battlespace and damaging forward air bases. One novel concept is the use of a low-cost attritable aircraft technology (LCAAT) class of weapon system that can be generated in large numbers without requiring a runway for combat power projection (however, a runway may still be needed for sustainment support). In an effort to take a holistic approach to future force design, AFWIC endeavors to evalu-

[1] See Alan J. Vick, *Air Base Attacks and Defensive Counters: Historical Lessons and Future Challenges*, Santa Monica, Calif.: RAND Corporation, RR-968-AF, 2015.

ate not just the operational effectiveness of new weapon system platforms but also what the employment of those weapon systems implies for force enablers like combat support (CS) and combat service support (CSS) resources.[2] This report will address the capabilities required to generate combat power away from fixed infrastructure using the LCAAT class of weapon system, with minimal reliance on a runway for CS or CSS.

The LCAAT class of weapon system is designed to fill a gap in the Air Force weapon and weapon systems portfolio of platforms. The Air Force recently announced that it seeks to develop a class of attritable platforms that range in cost between $2 million and $20 million. That makes the attritables more expensive than air-launched missiles and expendable unmanned aircraft that range from $100,000 to $2 million, and cheaper than conventional platforms that cost $20 million and up.[3]

As envisioned by the USAF, the LCAAT is a high-performance weapon delivery and sensor weapon system that can be launched from mobile platforms. The concept of launch using a variety of approaches that do not include a runway allows the USAF to deliver combat power if adversary attacks render an air base and runway inoperative. The idea is that mobile, low-cost platforms can be acquired and employed in large quantities, in a manner that leverages the principle of mass as a means to exhaust the adversary's quiver of air defense weapons. Likewise, by not having its weapon system tied to a runway, the USAF can leverage the principle of maneuver and maintain operational resilience in the face of attacks on air bases and runways.

---

[2] The U.S. Department of Defense defines CS as "[f]ire support and operational assistance provided to combat elements" and CSS as "the essential capabilities, functions, activities, and tasks necessary to sustain all elements of all operating forces in theater at all levels of warfare." U.S. Department of Defense, *DoD Dictionary of Military and Associated Terms*, June 2020. For our purposes, we include maintenance, fuel, and munitions activities in the category of CS and all other base operating support functions (e.g., security forces, transportation, engineering, lodging, messing) as CSS.

[3] Steve Trimble, "USAF Defines Price Range for 'Attritable' UAS," *Aviation Week*, August 3, 2020.

Prior RAND PAF research focused on the deployment and employment (i.e., mission generation) requirements associated with the LCAAT class of weapon system.[4] That analysis highlighted that the LCAAT can be deployed and operated using a fraction of the support footprint associated with conventional platforms. It noted that the employment of a version of an LCAAT, the XQ-58A, could be accomplished without dependence on a runway, though a runway may still be required to support sustainment operations. Hamilton and Ochmanek addressed the deployment and employment requirements of a smaller variant of the LCAAT class referred to as the L-Kitten. Neither focused on the broader implications of LCAAT operations for logistics support and sustainment requirements for mobile teams that could generate combat power without *any* reliance on a runway, either for combat force projection or logistics support and sustainment.

Research Questions and Approach

In this report, we build on the previous research and quantify the broader logistics support and sustainment requirements for LCAAT operations. We do not seek a point solution but rather explore the trade-offs associated with different capabilities and support concepts. Specifically, the fundamental research questions we sought to explore were

- What categories of logistics support will LCAAT units require?
- For each category of support, what are some alternatives for providing the required support?
- Are support alternatives likely to qualitatively affect operational resiliency?
- What are the deployment and employment requirements for alternative support concepts?
- How can LCAAT design modifications enable better logistics support concepts?

---

[4] See Leftwich, 2020; and Thomas Hamilton and David Ochmanek, *Operating Low-Cost, Reusable Unmanned Aerial Vehicles in Contested Environments: Preliminary Evaluation of Operational Concepts*, Santa Monica, Calif.: RAND Corporation, RR-4407-AF, 2020.

To answer these questions, we begin by establishing as a baseline the quantity of CS and CSS resources that must be deployed to establish combat operations for traditional weapon system platforms. We then decompose the functions of a traditional forward-operating location and identify which support functions are required for LCAAT operations and which are not. For the required support functions, we identify options for providing the necessary capability. We piece together options for individual support functions into a few overall support concepts that provide different levels of operational resilience and deployment and employment requirements. We developed a model of support concepts that enables this analysis. Finally, after identifying key drivers of support requirements, we examine design alternatives that could better enable the support concepts.

In exploring alternatives for different support functions, we used a "capability–concept of operations" structure, shown in Figure 1.1. We started with current capabilities in the USAF inventory and current concepts of operations. We then looked at other, nonmainstream USAF capabilities (e.g., capabilities used by specialized Air Force units

Figure 1.1
Approach to Analyzing Logistics Support to LCAAT Operations

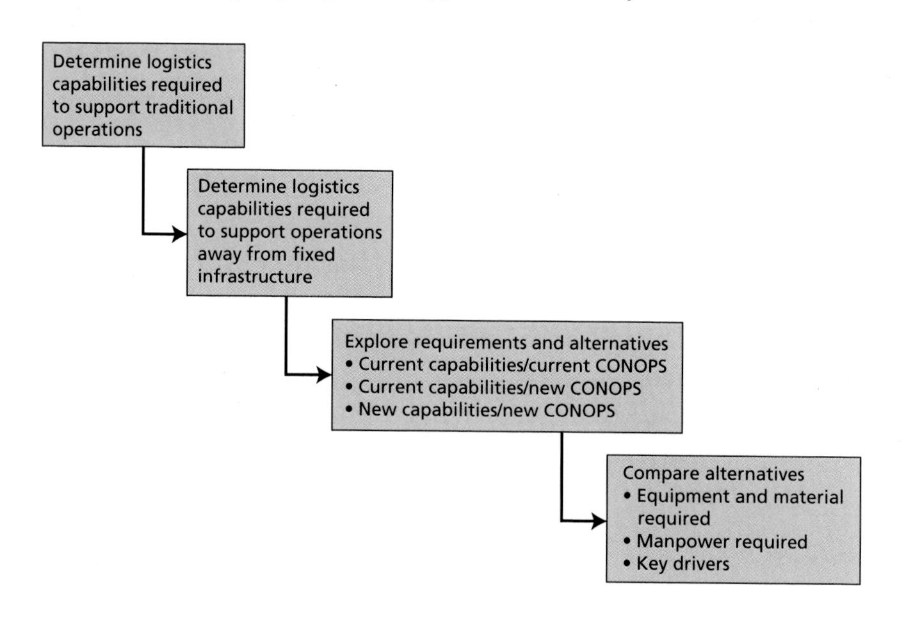

such as special operations and contingency response), as well as current capabilities that exist in other military services and commercially, and explored their use in new concepts of operation that are not common in the USAF. Finally, we considered new capabilities that could be achieved through engineering design modifications to the LCAAT and employed those using new concepts of operation. In all cases, we do not analyze every potential new capability and new concept of operations, but instead focus on a few exemplars to demonstrate how different capabilities and concepts could affect the overall equipment and manpower requirements.

To focus our analysis, we examined an instantiation of the LCAAT family of weapon system currently undergoing experimentation by the Air Force Research Laboratory (AFRL), the XQ-58A. While the XQ-58A served as the baseline for sortie generation requirements, fuel consumption, and weapon loadout, our analysis explores the ways in which other performance factors such as aircraft preparation and turn time, as well as manpower required for both, affect CS and CSS requirements.

Organization of This Report

The remainder of this report is structured as follows:

- Chapter Two establishes the baseline for operations at traditional forward operating locations (FOLs), decomposes that capability into a set of key functions, and identifies the functions required for LCAAT operations.
- Chapter Three introduces, generally, how those support functions are integrated with operations to form an increment of LCAAT capability.
- Chapter Four identifies specific options for each support function.
- Chapter Five pieces together individual options to form support cases and assess deployment and employment implications.
- Chapter Six describes system design options that could better enable LCAAT support.
- Chapter Seven presents observations and recommendations.

CHAPTER TWO

Determining Required LCAAT Support Functions

Traditionally, the Air Force conducts combat operations from forward air bases, some of which require significant buildup time, necessitate large amounts of real estate that become appealing targets to the adversary, and are vulnerable to enemy attacks. In this chapter, we identify the necessary support functions of the LCAAT class of platforms that are capable of operating without reliance on a runway and away from fixed infrastructure, though a staging base may be required for sustainment and logistical support.

Decomposing the Functions of a Traditional Forward Operating Location

To begin, we decompose the elements of an FOL used for conventional weapon systems. In doing this, we break out those capabilities and resources needed to support sortie generation, and those needed to support the base population, referred to as base operating support (BOS).

Sortie Generation Functions (Combat and Combat Support)

For conventional weapon systems platforms, several functions are required when operating from an FOL. Activities most directly tied to launching a sortie include flightline maintenance personnel and equipment, personnel and equipment to refuel the aircraft, and munitions personnel and equipment for loading weapons onto the platforms. These are the key direct sortie generation activities.

In addition to flightline maintenance activities, other repair activities must be performed to keep an aircraft mission-capable. These fall into two general categories and require both personnel and equipment: on-equipment maintenance (repair performed directly on the aircraft) and off-equipment maintenance (repair activities performed on a component of the aircraft after it has been removed from the aircraft). Additional fuel and munitions personnel and equipment are also required to maintain inventories of those resources at the FOL. For our analysis, we refer to these functions as fuel and munitions storage and distribution. Fuel storage and distribution in particular requires significant equipment and vehicles. Munitions can as well, but the quantities are dependent on the types of weapons being used.

From a safety perspective, additional functions include crash-rescue, fire suppression, and explosive ordnance disposal (EOD), the last of which is needed in the case of a munition being "stuck" (referred to as "hung") on the aircraft. For fire suppression and EOD, the personnel are dual-purposed at the FOL and perform their functions not just in support of sortie generation but also for the infrastructure on the installation (in the case of fire suppression) and unexploded ordnance on the installation that may have resulted from an adversary attack.

Finally, for mission generation, airfield operations functions are required. These include such things as air traffic control, airfield lighting, and aircraft arresting systems. There are also requirements for the flightline that include such items as barriers and security forces personnel that provide entry control and aircraft security.

Base Operating Support Functions (Combat Service Support)

As the name suggests, other personnel and equipment are required as a function of operating from a fixed infrastructure (i.e., a base), receiving forces at the location, and supporting the population of the installation. We break down the BOS functions into three categories: functions to open the base and receive forces, functions to establish and operate the base, and functions to repair the runway and base in the case of an adversary attack.

Opening the base and receiving forces requires force protection personnel that secure the installation. These personnel are generally above and beyond the security personnel that protect the flightline and aircraft. Personnel and equipment are also needed to establish communications within and outside the installation. Additionally, equipment and personnel are needed to offload any aircraft arriving to deliver units deploying from outside the theater and to distribute the arriving personnel and equipment to various operating locations around the base. Some of the equipment and vehicles (e.g., forklifts, buses, flatbed trucks) will subsequently be used to establish and operate the base.

The personnel and equipment required to establish and operate the base compose the majority of resources needed to operate an FOL. Personnel within the civil engineering career field provide heavy construction equipment for preparing the site (e.g., leveling the ground and building berms for fuel bladders) as well as the personnel and equipment to build lodging, feeding, and personal sanitation facilities. They also provide facilities for workplace operations (e.g., maintenance facilities and other office-type facilities), construct aircraft hangars, and establish the power and water infrastructure for the base. As mentioned earlier, the fire suppression and explosive ordnance personnel will often be dual-hatted, performing those functions for both mission generation and BOS.

Other functions required for operating a base include medical facilities and equipment, personnel accountability, force support (e.g., morale, welfare, and recreation), postal, expanded communications infrastructure, vehicle operations and maintenance, public affairs, legal, and contracting, to name a few. Many of these require minimal equipment but do add to the total base population that needs to be supported.

The final category of BOS is airfield recovery and runway repair. These capabilities are essential in a high-threat environment where a capable adversary can disable the operations of a base with missile strikes. For airfield recovery and runway repair, some of the construction equipment used to construct the facilities and prepare the berms can be used to support runway repair operations. Additional resources required for runway repair include other specialized heavy equipment and special construction materials such as quick-dry concrete.

Establishing a Baseline for Comparison

As the LCAAT likely will be used to deliver weapons as part of a hybrid force that consists of LCAAT and conventional weapon systems (e.g., MQ-9, F-35, F-16), we first established a baseline using conventional platforms against which to compare the LCAAT. We focused on weapons delivered as the operational demand signal. In this case, we posited a common demand signal of 800 naval strike missiles (NSMs) or 3,200 small-diameter bombs (SDBs) delivered in a 24-hour period as the required combat effect.

Drawing on prior RAND PAF research, Table 2.1 shows, for a conventional manned platform and an unmanned platform, the total number of aircraft required, the number of FOLs[1] required, and the total number of personnel and amount of equipment required to deliver the required weapons. Analysis in the prior study was conducted

Table 2.1
Beddown Requirements for Conventional Platforms Required to Deliver Desired Combat Effect

	F-16	MQ-9
Aircraft required	160	320
Number of bases[a]	5	9
Total personnel	8,736	10,519
Total STONs[b]	16,116	21,112

SOURCE: Adapted from Leftwich, 2020.

[a] We include the number of bases since it drives a large requirement for material and equipment. We limited the total number of aircraft at a base to a maximum of 36.

[b] Short ton (STON) estimates do *not* include airfield damage repair equipment, which would significantly increase footprint.

---

[1] Leftwich, 2020, used conservative estimates in determining the number of bases needed to bed down the required aircraft. In a theater with adversaries capable of striking bases with missiles, an approach to achieving survivability is to use a more dispersed beddown posture (i.e., fewer aircraft per base) requiring more locations than what is suggested here.

using a RAND PAF model called Lean–Strategic Tool for Analysis of Required Transportation (Lean-START).[2]

For the purpose of comparison later in the report, we will use the F-16, which compared most favorably in terms of STONs required. It is worth noting that the total STONs listed in Table 2.1 do not include any rapid airfield damage repair (RADR) assets that would be required to maintain operational resiliency if the airfield was attacked. The footprint for RADR kits starts at over 500 STONs for a small kit, over 1,400 STONs for a medium kit, and increase even more for large and very large kits.

LCAAT Operations and Required Support Functions

As highlighted earlier, the USAF envisions the LCAAT as a relatively high-performing platform, at a low cost, with an expected shorter life than conventional platforms. Original discussions with AFRL on the LCAAT class suggested the USAF was targeting platforms at the lower end of the cost range, roughly $2–$5 million per airframe.[3] The XQ-58A, a prototype of which is undergoing flight tests by AFRL, falls into that range. Given the availability of data and performance characteristics for this prototype, we used it as the baseline for our analysis.

The XQ-58A is modeled after USAF unmanned targeting drones that are capable of launching from a rail platform using strap-on rockets.[4] For the XQ-58A, the rail platform is replaced with a trailer, similar to a boat trailer, that makes the platform mobile. The value of the mobile trailer is that the XQ-58A does not require a runway to launch, thereby providing some degree of operational resiliency should

---

[2] See Patrick Mills, James A. Leftwich, Kristin Van Abel, and Jason Mastbaum, *Estimating Air Force Deployment Requirements for Lean Force Packages: A Methodology and Decision Support Tool Prototype*, Santa Monica, Calif.: RAND Corporation, RR-1855-AF, 2017.

[3] Discussion with Bill Baron and Doug Meador, AFLR XQ-58A program office, July 20, 2020.

[4] Discussions with AFRL revealed that the use of rockets to assist in takeoff is not the most desired approach. We address this later in the report, looking at the impact of rockets on the logistics footprint and some alternatives that could be considered.

an adversary attack the runway at an FOL, a key design objective of the XQ-58A. The added benefit of the launch platform is its maneuverability, making it more difficult for an adversary to target.

Current concepts portray the LCAATs being launched and integrated into an airborne data network composed of hundreds of aircraft, to include other LCAATs as well as conventional manned weapon systems. The LCAATs would need to be launched in high volume over a short period of time. Once the LCAAT completes its mission, it can return to any LCAAT operating area (in the case of multiple operating areas), not just the area from where it was launched. Current plans are for landing and recovery of the LCAAT to be accomplished by a parachute system (similar to the target drones mentioned earlier), again, making it runway independent.[5]

Upon returning to the launch area, the LCAAT will be prepared for its next mission. We break down the turn cycle in the next chapter and provide analysis of the impact of turn-cycle duration on the overall footprint. We note that as attritable systems, LCAATs will require little to no on- or off-equipment repair. Given the XQ-58A's modular design, cannibalization of operable modular components from multiple damaged aircraft in order to reassemble "new" mission-capable platforms can be accomplished by the same launch teams used to prep original aircraft for flight.

As we analyze the support requirements for LCAATs, we make a few assumptions:

- Many of the resources required for LCAAT operations are suitable for pre-positioning at forward locations (e.g., munitions, fuel storage and distribution equipment, vehicles, LCAAT operations equipment).
- There will be sufficient time building up to the onset of hostilities for actions needed to "set the theater."

---

[5] This is another area that is being examined by AFRL as they explore precision-guided recovery systems and other alternatives that do not require a parachute system. In our analysis, we explore the implications of the use of parachute landing systems and the value of alternative landing and recovery approaches.

- Staging bases will have use of a runway prior to commencement of combat operations, allowing for support personnel and LCAAT support and sustainment resources to be in place.
- Support and sustainment infrastructure for dispersed operations can be set in remote locations.
- Though the LCAAT version that we modeled can carry sensor payloads in lieu of weapons, we model weapons, given that it is likely the more logistically demanding case.

Mapping Key Functions to Be Modeled

In developing our model, we examined which of the key functions described earlier for conventional platforms would also be needed for LCAAT operations. Table 2.2 highlights the number of STONs required for an F-16 delivering the required 3,200 SDBs or 800 NSMs in a 24-hour period and the anticipated results for LCAAT operations

Table 2.2
Comparing Functions Required for Conventional Weapon Systems and LCAATs

Category/Function	F-16 (5 Bases)	LCAAT	
		Staging Base	Dispersed Ops
Sortie generation (combat and CS) *Operations/maintenance* *Fuel distribution* *Munitions distribution*	STONs: 2,751 Personnel: 5,130	STONs: 0 Personnel: 0	*The subject of our analysis in this report.*
BOS (CSS) Open base/receive forces *Establish/operate the base*	STONs: 13,365 Personnel: 3,606	STONs: 4,836 Personnel: 853	
Recovery (additional required for small RADR)	STONs: 2,535	STONs: 507	

NOTE: For recovery, we include (for information purposes only) the number of STONs required for airfield damage repair material and equipment and assume that each F-16 base would have, at a minimum, a small rapid RADR kit. We also include a small RADR for the staging base if the runway were to be used for delivering sustainment resources for LCAAT operations.

relative to the conventional operations. Those functions that are italicized became the foci of our LCAAT support analysis.

For the F-16, the USAF would need to establish five FOLs in order to bed down the number of F-16s required, and do so in a manner that does not mass too many aircraft at any single location. For our analysis, we assumed that three of the five locations would be low-capability bases, meaning very little support is available at those locations. We planned for the other two to be medium-capability bases, such as a collocated operating base or international airport.[6]

Our analysis in Chapter Five will fill in the numbers in the column for LCAAT operations in Table 2.2. We hypothesize that for mission generation, the support requirements for the LCAAT will be lower than those for the conventional platforms; this is due in large part to the focus on aircraft launch for the LCAAT, rather than on- and off-equipment repair. The fuel and munitions storage and distribution functions are two key areas of our analysis. The exemplar of the LCAAT that we are modeling requires less fuel than a conventional platform, but it also has a smaller weapon delivery capability. The munitions storage and distribution requirements should be generally the same. Since we are basing our analysis on combat effects delivery, the same number of weapons would need to be stored. Our analysis will, however, explore the impact of operating from dispersed locations beyond the perimeter of an air base.

Turning to BOS, and highlighted in Table 2.2, the first takeaway is that the total requirement for the F-16 is driven by the need to establish five different operating locations in order to bed down the 160 aircraft required. The resources required to open the staging base for the LCAAT should be the same as a single traditional FOL. For

---

[6] We defined a low-capability base as one that has a runway but little else in terms of infrastructure needed to support both combat power projection and a base population. The medium-capability base has a runway and more-robust infrastructure; however, the Air Force would not have access to all of the infrastructure, as it would likely be in use by the owning or hosting entity (e.g., coalition partner or commercial port authority). As a result, the Air Force would need to provide accommodations for the base population but would not need to provide airfield operations resources, such as navigation aids, arresting gear, or fuel storage, to name a few.

establishing and operating the base, our analysis examines alternatives that better support distributed operations and require less time to be constructed. Our primary focus related to base operations is on lodging, messing, and security. We offer a core element of personnel that would be needed for such items as medical support, communications, and air operations. Finally, the need for runway repair capability would be eliminated for LCAAT operations since the runway is not needed to launch the version of the LCAAT we are modeling.[7] Again, the focus of our analysis is on decoupling the generation of combat power from fixed infrastructure.

The BOS requirements for the staging base were determined using the Lean-START model mentioned earlier. Given that the staging base is the source of bulk supply for the LCAAT operations, and perhaps even the storage location for LCAAT attrition fill and other pre-positioned assets, we sized the requirements to the need to establish an operational base that could house and secure those activities. The BOS required for the staging base includes functions such as lodging and messing for personnel on the base, transportation, additional fuels storage and distribution, base security, fire and medical support, warehouse support, and general engineering support. If the runway were to be used to deliver sustainment resources for subsequent distribution to the LCAAT operations, the base would likely have airfield damage repair capability as well; however, as with the F-16 locations, we exclude that from our total requirement. Given the criticality of the staging base to LCAAT operations, we would expect that, similar to beddown locations hosting conventional platforms, the staging base would have some sort of active and passive missile defense capability. We also exclude those requirements from both the conventional aircraft bases and the LCAAT staging base.

For our analysis, we model the staging base as a single location to demonstrate the personnel, equipment, and material requirements. In reality, the LCAAT operations could be dispersed in cluster-size units

[7] For our analysis, we conservatively exclude the airfield damage repair resources for the F-16 bases in the equipment and material requirements. We include them in Table 2.2 for information purposes.

(discussed in the next chapter) and collocated at FOLs hosting conventional platforms. Such an approach would allow for the LCAATs to leverage the active and passive missile defense capabilities at those locations, and still allow for the execution of combat operations in the event of an adversary attack on the runway.

In the next chapter, we describe the basic concepts of employment and support for LCAAT operations.

CHAPTER THREE

Linking LCAAT Support Functions to Operational Concept

Before conducting an analysis of support options for LCAAT operations, we need to define an increment of LCAAT capability by generally describing how the LCAAT support functions in the previous chapter link with the operational concept for the LCAAT. Previous research defined an LCAAT unit as capable of launching 100 LCAAT sorties every three hours, providing a force projection capacity of 800 sorties (3,200 SDBs delivered) a day.[1] That definition was appropriate as a starting point to develop a first-order estimate of the manpower and launch equipment required to operate LCAATs. In this report, we use a more specific definition of LCAAT capability increments to (1) account for the inherent uncertainty around LCAAT turn times (e.g., a unit required to support 800 sorties a day given a three-hour turn time is likely much smaller than one with a six-hour turn time), (2) provide a more modularized concept such that capability can be provided in smaller increments (i.e., increments smaller than 800 sorties per day), and (3) enable definition of operations support concepts with varying degrees of centralization.

Rather than specifying an LCAAT unit in terms of the number of daily sorties, which varies significantly based on turn-time assumption, we will define a unit in terms of the number of launchers, and associated launch teams, it contains. We organize LCAAT capability into three levels:

[1] Leftwich, 2020.

- Launch site: an individual launcher and associated launch team (a launch team is composed of four personnel). A launch site can provide a certain number of sorties per day depending on the turn time. For example, if the turn time is three hours, a launch site provides eight sorties per day.
- Launch cell: a group of launchers and launch teams that share a recovery area and may or may not share other resources such as fuel and munitions storage or beddown support. A launch cell can provide a certain number of sorties per day depending on the number of launchers per cell. In our analysis, we assume six launchers per cell.
- Launch cluster: a group of launch cells that may share resources such as fuel and munitions storage and/or beddown support. The number of sorties provided by a cluster depends on the number of launch cells per cluster. In our analysis, we assume three launch cells per cluster.

Figure 3.1 depicts a notional launch cluster, composed of three launch cells and 18 launch sites. Each blue rectangle is a notional launch cell that contains six launch sites, represented as blue triangles, and a shared launch cell recovery area. Housing of personnel, fuel stor-

Figure 3.1
Notional Launch Cluster

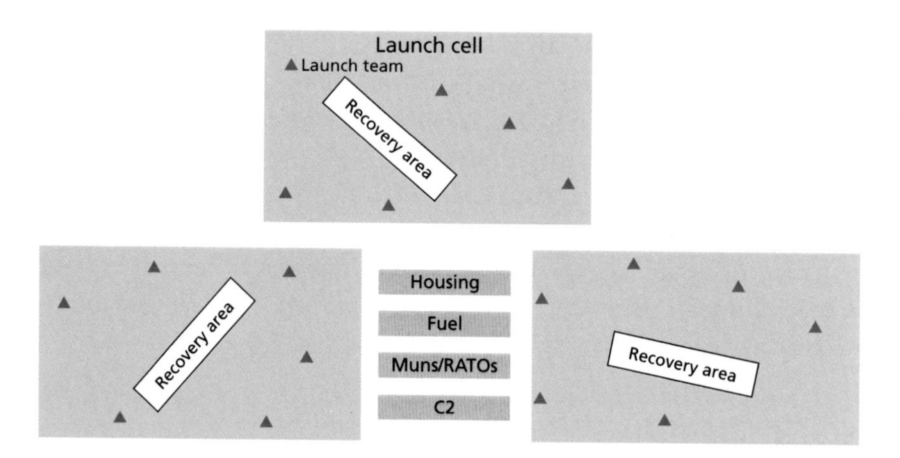

age, munitions storage, and command and control (C2) are required to support LCAAT operations. In Figure 3.1 they are shown as centralized to the launch cluster; however, these could be distributed out to the launch cells or even all the way out to the launch sites. These support options will be discussed more in the next chapter. Launch clusters are supported by a staging base (not shown in the figure) that serves as the central storage and supply for fuel and munitions. The number of launch teams per cell and the number of launch cells per cluster affect the potential for sharing or distributing housing, fuel, munitions storage, and C2. For example, a different amount of fuel storage would be required at the launch cell depending on how many launch teams it contains.

Implications of LCAAT Assembly and Turn Times

The sortie production of the launch cluster described above depends on the LCAAT assembly time and turn time. We define assembly time as the time it takes for a launch team consisting of four personnel to pull a previously unused LCAAT out of its storage container; assemble as needed; conduct initial system checks; load fuel, munitions, and rocket-assisted takeoffs (RATOs); and launch. The turn time is the time it takes a launch team to recover an LCAAT after its previous sortie; bring it back to the launch location; load fuel, munitions, and RATOs; and launch it again. According to our conversations with subject-matter experts (SMEs), estimates for assembly time and turn time vary greatly. Assembly times range from 4 to 12 or more hours, and turn times from 2 to 12 hours. If assembly time is less than or equal to turn time, then turn time drives the number of sorties a cluster can generate in a day. If assembly time is longer than turn time, and assuming some number of LCAATs will be lost to either combat or maintenance attrition, then both assembly time and turn time affect sortie production.

We first examine the simpler case, where assembly time is less than or equal to turn time. The number of sorties generated by a cluster is the number of launch sites times the number of sorties per launcher

Figure 3.2
Cluster Sortie Generation as a Function of Turn Time

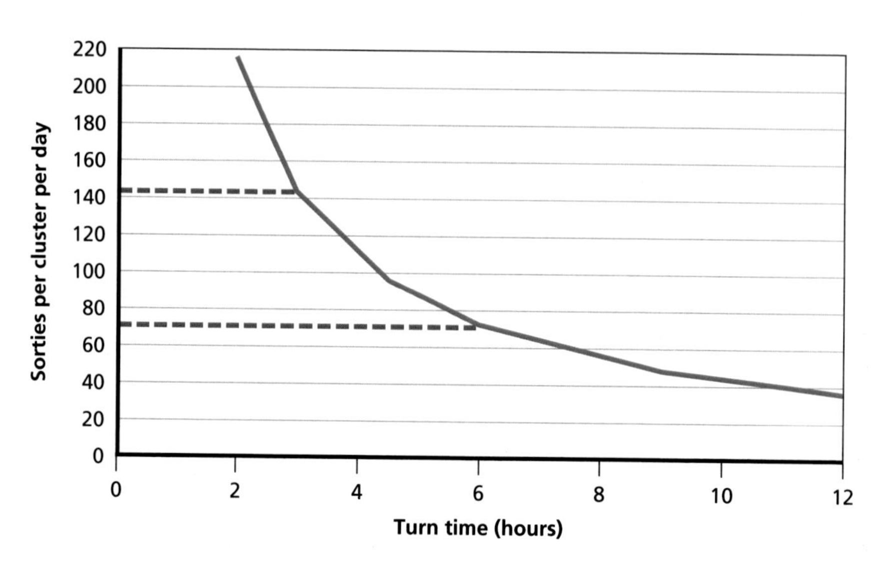

per day. Figure 3.2 shows the number of sorties generated per cluster as a function of turn time. If the turn time is three hours, then each launch site can generate eight sorties per day for a total cluster sortie generation of 144 (8 sorties × 18 launch sites). If the turn time is six hours, then each launch site can generate only four sorties per day for a total cluster sortie generation of 72 (4 sorties × 18 launch sites). For longer turn times, more clusters will be required to provide an equal number of sorties. Turn time, then, is a key driver of the total amount of support required.

Table 3.1 shows a task-by-task breakdown of the turn cycle with worst- and best-case estimates for the time it will take to conduct each task. As shown in the table, three tasks account for more than 70 percent of the total time in the worst case. The first two tasks, parachute recovery and parachute/airbag removal, involve finding the LCAAT after it lands, bringing it back to the launch site, and removing the landing system. Worst-case estimates indicate four hours total for these tasks, while best-case estimates indicate just over half an hour. Similarly, the landing system install is estimated at 6.5 hours in the worst

Table 3.1
LCAAT Turn-Cycle Task Breakdown

Task	Worst Case (hours)	Best Case (hours)
Parachute recovery	2	0.3
Chute/airbag removal	2	0.3
Mission download/Mx documentation	0.3	0.2
Inspect/repair/replace damage	1	0
Refuel/top off	0.5	0.5
Chute/airbag/nitrogen/PAD/CAD install	6.5	0.5
Weight and balance	0.3	0
Load to rail for launch	0.5	0.3
RATO upload	0.75	0.3
Supervisory release	0.1	0.1
Launch	0.5	0.1
Total	**14.45**	**2.6**

SOURCE: Information supplied to the authors by AFWIC.

NOTE: Worst-case estimates are based on demonstrated capability of the XQ-58A assuming no improvements are possible. Best-case estimates are based on objective goals derived from system modifications.

case and just 30 minutes in the best case. Reducing the uncertainty in the estimates for these tasks and/or designing the system to achieve as close to the best-case estimates as possible will be important for understanding and maximizing the capability of LCAAT operations.

Assuming that LCAAT launch cycles are set based on turn time, if the assembly time is longer than the turn time, then a launch cycle would be interrupted for every LCAAT lost to attrition (e.g., if the assembly time is eight hours and turn time is six hours, a new LCAAT could not be assembled in time to meet the next launch). In this case, it may be desirable to pre-assemble LCAATs before the start of combat operations. The number of LCAATs that would need to be pre-assembled

Figure 3.3
Required Pre-Assembly Time

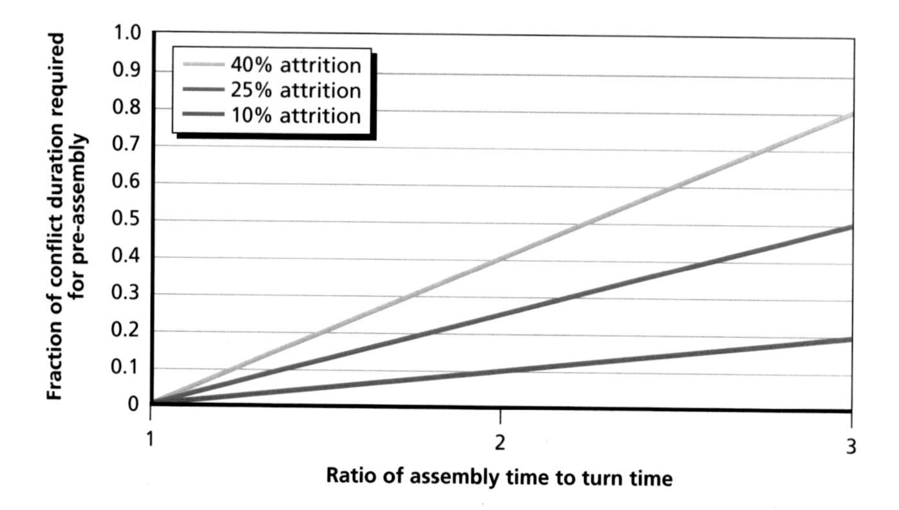

would depend on the duration of the expected conflict and the anticipated level of attrition. The amount of pre-assembly time needed would depend on that number of LCAATs and the pre-assembly time. Figure 3.3 shows the fraction of conflict duration required for pre-assembly as a function of the ratio of assembly time to turn time. When assembly time is equal to turn time, the ratio is 1, and no pre-assembly time is required. As the assembly time becomes longer than the turn time, the time required for pre-assembly increases. For any given ratio of assembly time to turn time, a higher attrition rate requires longer total pre-assembly time.

Depending on the amount of strategic warning, and when the equipment and personnel arrive at the operating areas prior to the onset of hostilities, there may not be sufficient time for pre-assembly. In that case, if the assembly time is longer than the turn time, we would expect a degrade in sortie production. Figure 3.4 shows the degrade in sortie generation as a function of the ratio of assembly time to turn time. As assembly time becomes longer relative to turn time (ratio increasing), total sortie generation decreases. For a given ratio of assembly time to turn time, higher attrition rates result in lower sortie generation.

Figure 3.4
Sortie Generation Degrades Without Sufficient Pre-Assembly Time

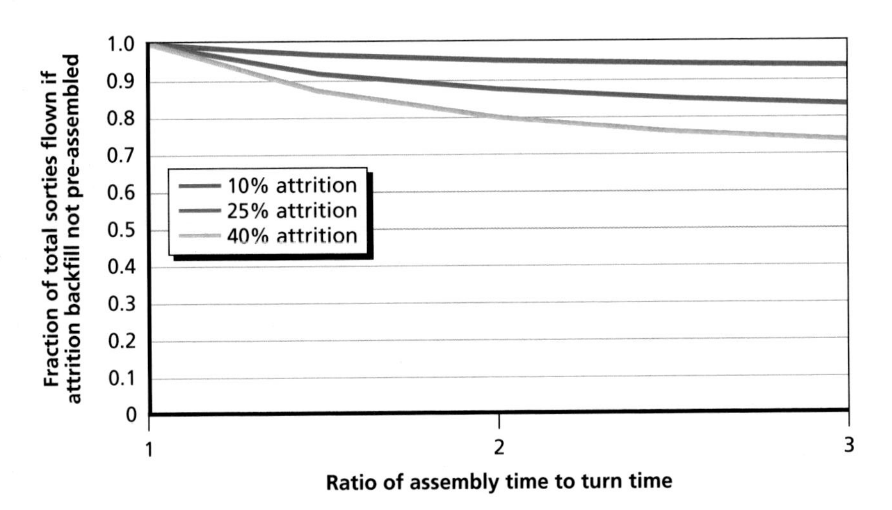

CHAPTER FOUR

Identifying LCAAT Support Options

Supporting the LCAAT operations described in Chapter Three requires three primary categories of support: fuel, munitions, and housekeeping. For each category, there are various ways to provide the required support. In this chapter and the next, we examine capabilities and tactics, techniques, and procedures used by specialized units in the USAF (e.g., special operations units, contingency response groups), as well as existing capabilities available in other services and commercially that could be applied to the LCAAT concept.

Fuel Support for LCAAT Operations

Depending on the size of the LCAAT, the fuel efficiency of the engine, and the fuel capacity, a maximum duration sortie will likely consume approximately 300–500 gallons of fuel.[1] Depending on the turn time, fuel consumption per launch site could range from 600 gal/day (12-hour turn time, 300 gal/sortie) to 6,000 gal/day (2-hour turn time, 500 gal/sortie). However, for a fixed amount of combat power, in terms of a set number of sorties, the total fuel requirement will be the same regardless of turn time; it will just be spread across a smaller or larger

[1] Fuel consumption estimates are based on RAND analysis informed by XQ-58A operating characteristics taken from the LCAAT presentation by Bill Baron, AFRL. See William Baron and Douglas Meador Baron, "Low Cost Attritable Aircraft Technology Initiative: XQ-58A Information for CONOPS Analysis," briefing slides, Dayton, Ohio: Air Force Research Laboratory, October 29, 2018.

number of launch sites. The fuel support for the operations will require fuel storage, potentially centralized or distributed to varying degrees, and fuel distribution via fuel trucks.

We considered fuel storage options ranging from small fuel bladders up to the very large fuel storage systems, as described here:

- 500-gallon containerized blivets.[2] Based on the 500-gallon bladders in hardened containers designed to support special operation forces (SOF) forces (unit type code [UTC] 4F9J4), we assume that a containerized fuel bladder could be built into the back of the LCAAT tow vehicle. The blivet weighs approximately half a ton and costs $30,000.
- Helicopter expedient refueling system (HERS).[3] The HERS is designed for refueling U.S. Marine Corps helicopters in support of operations in forward or remote locations, which is an operation analogous to the forward refueling required for LCAATs. HERS has a capacity up to 18,000 gallons, which is scalable using combinations of 500- or 3,000-gallon tanks. The full 18,000-gallon configuration weighs 3.1 tons and costs approximately $100,000.
- Fuels operational readiness capability equipment (FORCE).[4] FORCE is the standard USAF equipment set that provides the entire range of fuel support for contingency operations. It provides large storage capacity scalable in 50,000-gallon increments. The full equipment set weighs 63 tons and costs $1.1 million.
- Joint offshore fuel farm (JOFF).[5] JOFF is a future concept for submerged fuel storage capability. The JOFF is scalable from 10,000 to 1 million gallons with costs estimated from $500,000 to $18 million depending on size.

---

[2] Chris Clauser, "Air Rapid Response Kit (ARRK) Capabilities Brief," Air Force Special Operations Command, undated.

[3] United States Air Force Pamphlet (AFPAM) 23-221, *Materiel Management—Fuels Logistics Planning*, March 11, 2013.

[4] AFPAM 23-221, 2013.

[5] Laine F. Krat, "Air Force Design/Futures Game 2019: Basing and Logistics Seminar—ACS Concepts," briefing slides, San Antonio, Tex.[0]: Air Force Installation and Mission Support Center, January 11, 2019.

Fuel storage to support LCAAT operations will likely be some combination of these options. Regardless of specific choices for fuel storage, there will be a requirement to transport fuel between a central storage location and the launch clusters, cells, and/or launch sites. We considered a variety of fuel trucks that could be employed to meet this need:

- R-11 fuel truck.[6] This is the most common fuel truck in the USAF inventory; it has a 6,000-gallon capacity. A limitation of this truck is that it is not off-road capable. One truck weighs 14 tons and costs approximately $200,000.
- C300 fuel truck.[7] This fuel truck has a 1,200-gallon capacity and is also in the existing USAF inventory. One truck weighs 7.3 tons and costs approximately $33,000.
- Large-capacity refueler.[8] This refueling truck has a capacity of 17,500 gallons, and it is currently in limited use by the USAF. Limitations are its ability to be moved by air, and restrictions on driving on prepared surfaces because of its size and weight. This vehicle is not off-road capable. One truck weighs 16,500 tons and costs $705,000.
- Heavy expanded mobility tactical truck (HEMTT).[9] The HEMTT is designed to transport 2,500 gallons of fuel to forward, austere locations. One truck weighs 21 tons and costs approximately $200,000.

Each truck type has advantages and disadvantages. The R-11 is a good compromise between capacity and weight and is likely the easiest for the USAF to obtain, given that it is the standard current refueler. However, it may not be suitable to operate in austere environments that may be required for LCAAT operations. Larger-capacity trucks,

---

[6] AFPAM 23-221, 2013.

[7] AFPAM 23-221, 2013.

[8] See Rampmaster Corporation, "17,500 Gallon WD Modular Lift Deck Jet Refueler," undated.

[9] AFPAM 23-221, 2013.

like the large-capacity refueler, would minimize the number of trips required between the central storage location and the launch clusters. However, these also may not be suitable for use in austere conditions. The HEMTT addresses the need for austere use but is heavier and has a smaller capacity than the R-11s, requiring more trucks and a larger deployment burden.

Munitions Support for LCAAT Operations

The primary employment concept for the LCAAT is to prosecute targets with either air-to-air or air-to-ground munitions. The anticipated air-to-air loadout would include four AIM-120 (335 lbs each) or similar munitions. Air-to-ground loadouts could be four SDBs (285 lbs each) or one NSM (900 lbs). Munitions support requires munitions storage, distribution, and handling equipment. For our analysis, we focused on the SDB.[10]

The general munitions support concept involves a central storage location(s) for munitions, likely a munitions storage area on an existing base with proper munitions storage facilities. We assume that the munitions are located in theater prior to conflict. On order, before the commencement of hostilities, munitions would be distributed from the central storage area to the launch clusters, cells, or sites via flatbeds towed by trucks.[11] We considered two options for munitions storage in the field: open storage on flatbeds and expeditionary storage pads using HESCO MIL barriers. Forklifts would be affixed to the flatbeds to enable unloading of munitions crates. We considered four alternatives for munitions loading:

---

[10] We chose the SDB because the current design of the XQ-58A accommodates the loadout of SDBs. The design of the XQ-58A would need to be changed to accommodate an NSM. We also viewed the SDB to be the more demanding case logistically.

[11] We address the personnel required to support the distribution of munitions later in this chapter when we discuss the additional CS and CSS requirements for LCAAT operations. Additionally, our speed of movement related to sustainment and resupply accounts for the fact that explosives are being moved.

- MJ-1. The MJ-1 is currently in the USAF inventory and is the standard lift truck used to transport, load, and unload a wide variety of munitions up to 3,000 pounds. It weighs two tons and costs $143,000.
- Manually operated lift truck (MOLT). The MOLT is in the USAF inventory and is typically used to load bombs downrange if no fuel or other powered lift trucks are available. It can lift munitions up to 2,450 pounds, weighs just under one ton, and costs approximately $23,000.
- Kubota (or similar) tractor. A potentially attractive option for an LCAAT operation would combine the mechanized lift of the MJ-1 with the austere operating ability of the MOLT. One potential option would be a slightly modified off-road tractor. We used the Kubota tractor as a surrogate. This tractor could provide off-road-capable bomb lift to support approximately 2,000 pounds, weighs just over one ton, and costs approximately $20,000.
- Hand or "hernia bar." The most labor-intensive method, but also the lightest and cheapest, would be to have maintainers load the munitions by hand with a hernia bar.

Beddown Support for LCAAT Operations

The LCAAT clusters are likely to be established in austere locations with no access to existing infrastructure for housekeeping, food, or water. We examined four potential housekeeping options:

- Basic expeditionary airfield resources (BEAR).[12] BEAR is the standard equipment set used by the USAF to establish a bare base capability in an austere operating location. The sets include bare base equipment such as shelters, generators, laundry, and shower/shave units to support up to 550 personnel. This standard equipment set weighs 395 tons and costs $5.3 million. It requires approximately 92 engineering personnel and five days to construct BEAR

[12] United States Air Force Handbook 10-222, Volume 2, *Bare Base Assets*, February 6, 2012.

lodging and messing facilities at an established airfield.[13] In the scenarios we analyze later, we assume these engineers would be drawn from the engineers deployed to the staging base.

- Air rapid response kit (ARRK).[14] The ARRK is a rapidly deployable force support kit used primarily by SOF designed for deployment in increments (UTC modules) that provides absolute minimums meeting specific mission needs. It provides basic shelter, sanitation, and support for 100 personnel (or 200 personnel if "hot bunking"). The standard set also includes C2 facilities, environmental control, and water purification. The standard set weighs 25 tons and costs $1.8 million. The ARRK requires a team of eight civil engineers to construct.[15]
- ARRK Lite. The ARRK equipment sets are modular by design. We considered an alternative that did not include the equipment necessary for C2 but houses the same number of personnel. This reduced the weight of the equipment set to support 100 people from 25 tons to 13 tons and the cost from $1.8 million to $1.1 million.
- Marine combat tents. The most primitive option we considered is the marine combat tent. The weight of tents to house 100 people is about half a ton, and their cost is $39,000. This option does not include an environmental control or sanitation. Personnel would carry and construct their own tents.

In all cases we assume three ready-to-eat meals per person per day, which would be shipped and stored in pallets (3,456 meals per pallet), and 19 liters of water per person per day,[16] which would be shipped and stored in pallets (4,320 rockets per pallet).

---

[13] Richard Varden, "PACAF Civil Engineer Force Modeling Adaptive Basing: Minimum Endurance," PowerPoint presentation, Joint Base Pearl Harbor Hickam, Honolulu, Hawai'i: Headquarters, Pacific Air Forces, October 4, 2016. Not available to the general public.

[14] Clauser, undated.

[15] For our analysis, we assume the same team of eight could erect multiple ARRKs over a one-to-two-day time period.

[16] United States Air Force Tactics, Techniques, and Procedures 3-34.1, *Services Contingency Beddown and Sustainment*, November 1, 2007.

To support 550 people, moving from the standard BEAR kit to the ARRK reduces the lift demand by more than half. In addition, the six ARRK kits could be distributed closer to the launch sites, reducing vulnerability to missile attack. Moving from the standard ARRK to the ARRK Lite reduces the lift requirement by half again. Moving to the marine combat tent reduces the lift requirement by another order of magnitude and enables even wider dispersal. However, for operations that are expected to last more than a few days, the austere living conditions in tents with no sanitation or environmental control may prove untenable.

Additional Combat Support and Combat Service Support for LCAAT Operations

The LCAAT operation will likely be conducted "outside the wire" or not on an existing air base. As a result, it would not benefit from the standard security provided as BOS and would require some additional CS and CSS.[17]

The LCAAT cluster CS supervision team provides one shift (12 hours) leadership (1), maintenance supervision (1), and production superintendent (4), weapon safety (4), and petroleum, oil, lubricants (POL) supervision (2) for an LCAAT cluster operating 18 LCAAT launch vehicles, two weapon storage areas, and a POL distribution point.

The LCAAT cluster CSS provides two shift (24 hours) support to an LCAAT cluster operating 18 LCAAT launch vehicles, two weapon storage areas, one POL distribution point, and a vertical takeoff and landing zone/medical evacuation point. The functions and personnel for supporting the cluster include force protection (6x security forces and 2x EOD), communication (1x radio frequency [RF] trans/1x cyber trans), C2 (1x command post controller), Mx (1x aerospace ground equipment or vehicle mechanic [VM] for small engine/equipment

[17] Information provided by SMEs that have been active in planning support to USAF Agile Combat Employment operations.

repair), medical (4x independent duty medical technicians, physician's assistants, or flight surgeons), and airfield ops (1x airfield management with landing zone safety officer certification).

Finally, this additional support includes the trucks and 40-ft flatbed trailers required for distribution and two security sport utility vehicle escorts for convoys bringing fuel, munitions, or resupply to and from the central storage location.

CHAPTER FIVE

Assessing LCAAT Support Cases

LCAAT Logistics Support Model

To assess the requirements for the LCAAT, we developed a model that calculates personnel, deployment, and sustainment requirements based on a set of operational and support option inputs. While the model itself is platform agnostic relative to the specific platform in the LCAAT class of weapon systems, we use performance data for the XQ-58A provided by AFRL in conducting our analysis. Figure 5.1 shows the basic elements of the spreadsheet model.

The model allows the user to define the operational scenario by specifying the days of operations and number of sorties required per day. The user may also specify certain mission parameters as indicated under "Mission Planning" on the left-hand side of Figure 5.1. The baseline model is set to the platform characteristics of the XQ-58A; however, the user can change the performance characteristics as needed to represent other instantiations of the LCAAT envisioned by the USAF. The final operational inputs focus on the force laydown plan for the LCAAT operating teams.

The second set of user inputs is tied to the concept for logistics support and sustainment. Referring back to Table 2.1, we focused on the concepts for fuel and munitions distribution and storage, as well as those BOS functions that are required for LCAAT operations. The model allows the user to select different alternatives for each of the logistics support options. The approach to considering the alternatives is discussed in the next section.

Figure 5.1
Basic Design of the LCAAT Logistics Model

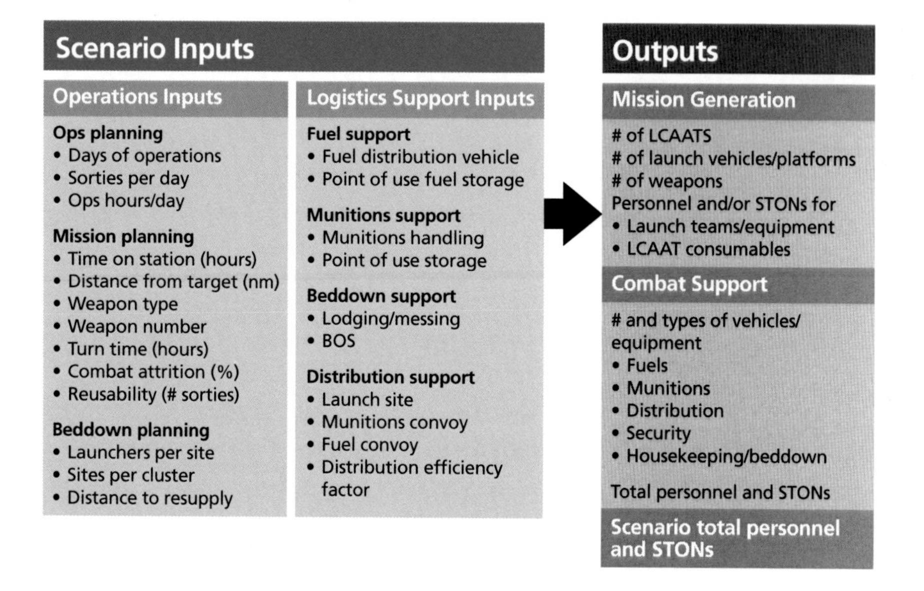

The far right side of Figure 5.1 shows the outputs for the model. From the operational scenario parameters selected, the model computes the number of launch teams and platforms required, the number of weapons required, and the total personnel and equipment footprint for the mission generation function. For the CS functions, the model computes both the equipment and personnel footprint required to provide fuel and munitions management and BOS.

Although the model allows for variation in operational requirements, our analysis focused on the requirements to support at least 3,200 SDBs per day, as discussed in Chapter Two. An LCAAT is capable of carrying four SDBs per sortie, so this equates to 800 sorties per day. To produce 800 sorties a day for a launch cluster that contains 18 launch sites and a turn time of three hours requires 5.6 launch clusters. To avoid fractional clusters, we round up to six clusters, which can produce 864 sorties, or 3,456 SDB strikes per day.[1] We calculate the

[1] Since this is slightly higher than the 3,200 SDB requirement for the F-16 case, the LCAAT estimates will be conservative.

equipment requirements to achieve this level of capability, which will vary depending on the support cases described above. We also calculate the sustainment requirements, which do not depend on support cases.

LCAAT Support Cases

The LCAAT operational concept is being developed based on demonstrated vulnerabilities to traditional basing concepts. Operating without a runway removes one of the primary vulnerabilities. However, in air base vulnerability studies, fuel and munitions storage and lodging have also been key vulnerabilities, and attacks on these have resulted in degraded sortie generation in the modeling.[2] At the same time, new employment concepts such as Agile Combat Employment have emphasized the need for small footprint capabilities to enable rapid deployment.[3]

To address operational resiliency, we focus on dispersal as the primary mechanism to achieve greater resiliency. However, in designing a support system for LCAAT operations, the aforementioned objectives (i.e., reducing vulnerability and reducing footprint) may not always align. More dispersed operations may reduce vulnerability, while centralized operations may provide for more resource sharing and, as a result, a smaller footprint. To explore these trade-offs, we consider three LCAAT support concepts with varying levels of dispersed operations and assess the footprint for each. We compare these with a fourth alternative that includes only options that are currently standard for USAF. In each case, resources are distributed to launch clusters from a central staging base before the start of hostilities. The four cases are the following:

- **Minimal dispersal:** fuel storage (HERS), munitions handling (tractors), and housekeeping (ARRK lite) resources are centralized at LCAAT clusters.

---

[2] Vick, 2015.

[3] Charles Q. Brown, Jr., *Agile Combat Employment (ACE): PACAF Annex to Department of the Air Force Adaptive Operations in Contested Environments*, Honolulu, Hawaii: Headquarters Pacific Air Forces, June 2020.

- **Medium dispersal:** fuel storage (HERS), munitions handling (tractors), and housekeeping (ARRK lite) resources are centralized at LCAAT cells.
- **Maximum dispersal:** fuel storage (HERS), munitions handling (tractors), and housekeeping (combat tents) resources are distributed to LCAAT launch sites.
- **Standard.** In this concept, we use only traditional USAF capabilities: FORCE, MJ-1, BEAR. Given the size of FORCE and BEAR capabilities, this will be a relatively centralized operation.

Personnel and Equipment Requirements for LCAAT Support

We used the LCAAT Logistics Support Model to calculate the number of personnel and the equipment required for each support case from the previous section. For the LCAAT cases we account for the personnel and equipment associated with both the support for operations as described previously and the staging base that would support the launch clusters. Table 5.1 shows the personnel requirements to support the conventional F-16 platform compared with the LCAAT operation when both approaches are sized to deliver at least 800 NSMs or 3,200 SDBs per day. The LCAAT provides the same operational capability with approximately 20 percent of the personnel required by the F-16.

Table 5.1
Comparing Personnel Needed to Support Conventional Versus LCAAT Operations

Category/Function	F-16	LCAAT
Sortie generation (combat and CS)	5,130	864
BOS (CSS)	3,606	246
Staging base (LCAAT only)	—	853
Total	**8,736**	**1,963**

NOTE: The F-16 numbers are from Table 2.2. Personnel estimates do not account for any attrition.

The number of personnel required does not vary by support concept, so we show just the single LCAAT case.

Table 5.2 compares the amount of equipment required for the F-16 and the four LCAAT support cases. For the F-16 case we include a small RADR set at each of the operating locations. In the LCAAT cases, we add the equipment and one small RADR for the staging base, and we also add the number of initial platforms to the equipment required. The initial number of LCAAT platforms required depends on the turn time and mission duration. If the mission duration is less than the turn time, then the initial number of LCAAT platforms required will be twice the number of launch teams. If the mission duration is more than the turn time but less than twice the turn time, then the number of required platforms will be three times the number of launch teams, and so on.[4] The number of launch teams required to

Table 5.2
Comparing Equipment Needed to Support Conventional Versus LCAAT Operations

		LCAAT			
Category/Function	**F-16 (STONs)**	**Standard (STONs)**	**Min Dispersal (STONs)**	**Med Dispersal (STONs)**	**Max Dispersal (STONs)**
Sortie generation (combat and CS)	2,751	2,159	1,933	2,052	2,311
BOS (CSS)	13,365	474	155	317	110
RADR	2,535	507	507	507	507
Staging base (LCAAT only)	—	4,836	4,836	4,836	4,836
LCAAT platforms	—	405	405	405	405
Total	**18,651**	**8,381**	**7,836**	**8,117**	**8,169**

[4] For example, if the turn time is three hours and the mission duration is two hours, then each team can launch the first sortie at time zero, would need a second platform to launch at hour 0+3, and then can relaunch the first platform at time 0+6. If the turn time is four hours, then the first platform is needed at time zero, a second at 0+3, and a third at 0+6, before the relaunch of the first is possible.

support the 864 sortie operation is 108, assuming a three-hour turn time. The fuel capacity of the LCAAT does not permit missions longer than seven hours, so the initial number of LCAATs required ranges between 216 and 432 platforms. At 1.25 STONs per platform, this results in an additional 270–540 STONs. In Table 5.2 our comparison is based on the midpoint of 324 LCAATs or 405 STONs.

In total, LCAAT operations require less than 50 percent of the materiel compared with the F-16. Less reliance on fixed infrastructure reduces vulnerability as well. A second finding is that moving from standard USAF assets to employing creative alternatives (e.g., Marine Corps mobile refueling system, USAF special operations housekeeping sets) for providing fuel, munitions, and beddown support can further reduce vulnerability by allowing more distributed operations for approximately the same footprint, as shown in the dispersal cases in Table 5.2.

In Table 5.2, sortie generation for the LCAAT includes the weight of the launch trailers, tow vehicles, and general-purpose vehicles as well as fuel support and munitions support. BOS includes beddown support and additional CS as described in Chapter Four. The sortie generation and BOS (first and second rows of Table 5.2) breakdown by specific function is shown in Figure 5.2.

As mentioned, LCAAT operational support includes the weight of the launch trailers, tow vehicles, and general-purpose vehicles required to support 864 daily launches. The number of these vehicles does not vary depending on support cases, and therefore the total weight associated with operational support is constant across the cases analyzed.

The total fuel required to support this level of combat power is also constant. However, the type and amount of fuel storage and the number of fuel trucks required varies depending on support case. For fuel storage, we assume that there must be adequate storage for at least three days of operations. The total daily fuel requirement is 411,000 gallons, so each case requires at least 1,233,000 gallons of storage. In the standard case, this required six FORCE sets, one per cluster. In the other cases, fuel storage was provided by HERS. HERS is configurable in 500-gallon increments, so roughly the same number of HERS is needed in each case. The maximum dispersal case, however, requires

Figure 5.2
Sortie Generation and BOS Equipment Requirement to Provide 864 Daily LCAAT Sorties

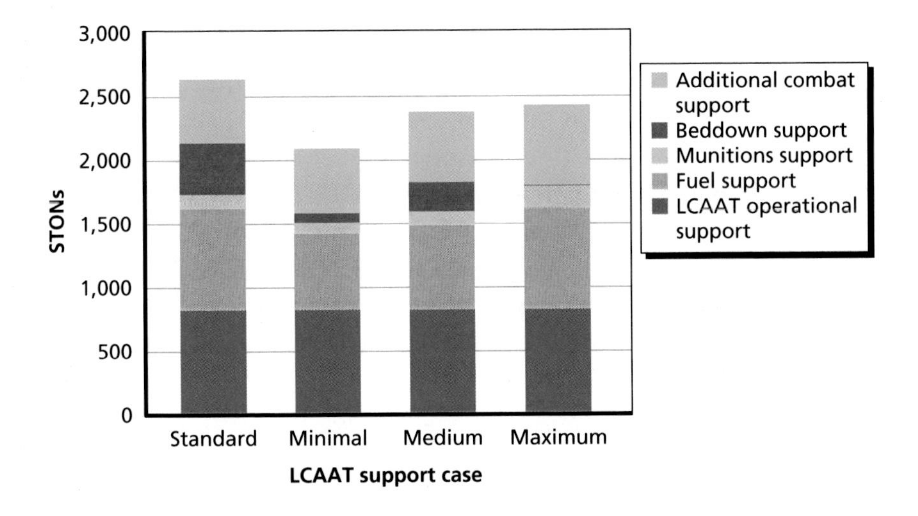

a somewhat larger number of HERS because they will be in substantially more locations. In all cases we assumed fuel distribution would be provided by the 2,500-gallon truck. We assumed that more truck hours would be required to distribute fuel all the way to the launch sites (rather than the tow vehicle bringing the air vehicle to some central fuel storage), so the number of trucks required increases as distribution increases (from 18 trucks in the minimal case to 25 trucks in the maximum case).

For munitions support, we assumed that munitions would be stored in the open at the cluster, cell, or site and would therefore not drive any significant differences in footprint. For munitions handling, we assumed that one MJ-1 would be required for every two launch sites in the standard case. For the other three cases, we assumed that the Kubota-like tractor would be used; the number required depends on the level of dispersion. In the minimal case we assume one tractor for every three launch sites; in the medium case, one tractor for every two launch sites; and in the maximum case, one tractor per launch site.[5]

[5] In each case we assumed the turn time would not be limited by tractor availability.

As a result, the munitions handling footprint increases as the level of dispersion increases.

Beddown support had the largest variance across the cases considered. Approximately 1,100 personnel are required to provide the 864 sorties per day, assuming the three-hour turn time. Assuming a two-shift operation and the ability to "hot bunk," this requires one BEAR set in the standard case. In the minimal dispersal case, six ARRK lite sets are required, one per cluster, significantly reducing the overall footprint. In the medium dispersal case, 18 ARRK lite sets are required, one per cell. This increases the footprint required to bed down the same number of personnel. Each ARRK lite set can house 200 people. When personnel are centralized at the cluster, as in the minimal case, the ARRK sets are nearly fully utilized. When distributed to the cells, the ARRK sets are substantially underutilized (i.e., the capacity of 18 ARRK sets is much greater than 1,100 people). In the maximum dispersal case, therefore, Marine combat tents are used as beddown support. As a result, the weight is nearly negligible. However, in this case no environmental control or sanitation is provided, and thus extended operations may be untenable.

Additional CS includes the trucks and flatbeds required to distribute the munitions, RATOs, and parachute landing systems from central storage out to the launch clusters, cells, or sites. We assumed more truck hours would be consumed if delivering to individual sites rather than central storage areas. To determine the number of trucks required, we constructed notional flatbed loadout configurations for transporting consumables. We calculated the number of trips required per day to supply one day's worth of consumables and subsequently calculated the number of trucks required. A notional flatbed loadout is shown in Figure 5.3. For the cases above, the number of trucks varied from 8 (minimum distribution) to 12 (maximum distribution).

In general, the support options affect the total equipment requirement only at the margins. We constructed a minimum case in which we selected the lightest option in each category. The total amount of equipment required across the support categories was 1,700 tons, a reduction of approximately one-third from the heaviest option. A more

Figure 5.3
Notional 40-ft Flatbed Loadout

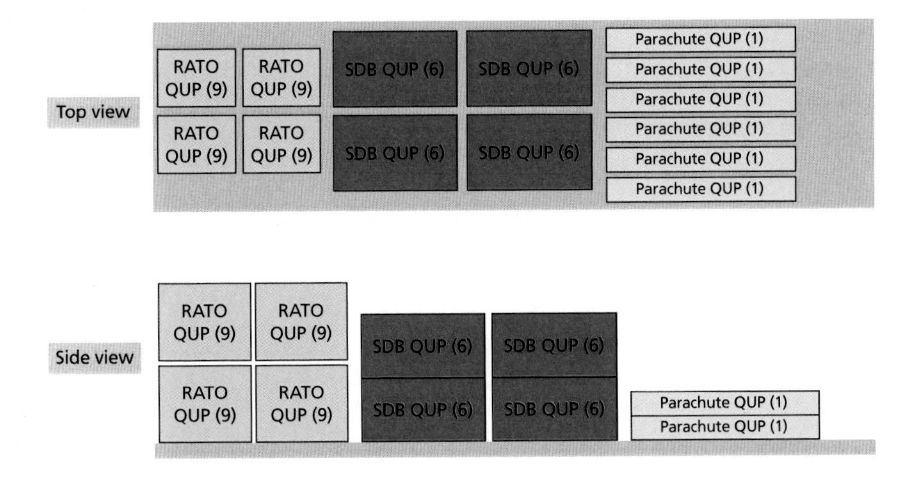

significant driver of requirements is the number of clusters required. The personnel and equipment in Tables 5.1 and 5.2 were the requirement for 864 sorties, assuming a three-hour turn time, which required six clusters. Figure 5.4 shows the personnel and equipment requirements for the "Medium Dispersion" option as a function of turn time. Longer turn times require additional launch sites to maintain the same combat capability. Doubling the turn time doubles the number of launch sites required. As a result, the personnel and equipment requirements increase significantly with turn time. Some categories of support, like LCAAT operational support and beddown support, increase proportionally with turn time, since those support categories are based on the number of launch sites. Other types of support, like fuel and distribution, increase only marginally since the same amount of fuel or munitions is being distributed, just to more locations. Given the sensitivity of support requirements to turn time, we explore alternative system designs in Chapter Six that could reduce turn time.

Another potential driver of equipment requirements is the travel time between the staging base (resupply location) and the operating locations. In the analysis presented earlier, we assumed a one-hour

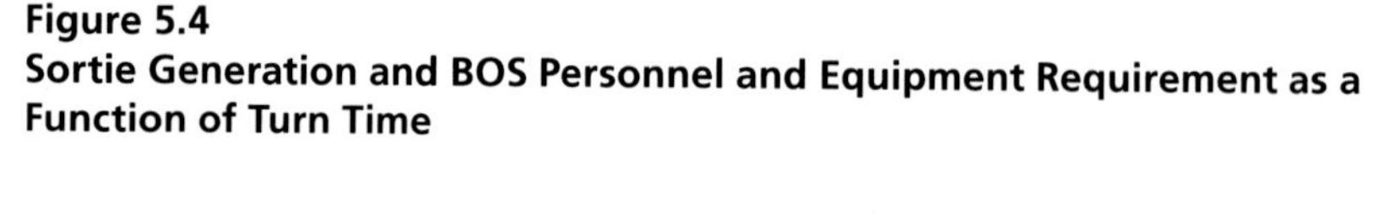

Figure 5.4
Sortie Generation and BOS Personnel and Equipment Requirement as a Function of Turn Time

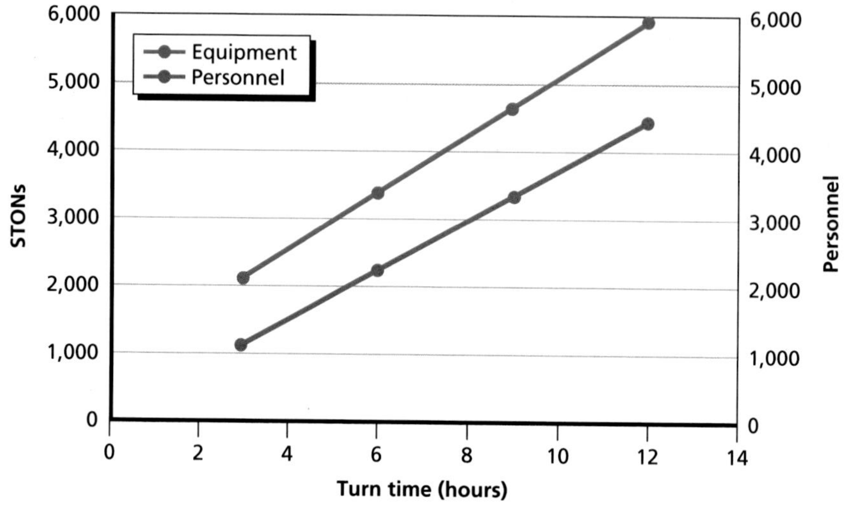

travel time. Given the number of trips between the staging base and operating locations required to deliver fuel, munitions, RATOs, and parachute/landing systems, as the travel time increases, the number of fuel trucks and distribution trucks and flatbeds will increase as well. Figure 5.5 shows the equipment requirements as a function of the travel time between the staging base and operating locations. If the travel time doubles from one hour to two hours, the additional trucks required results in an overall increase of about 25 percent.

Sustainment Requirements for LCAAT Support

The sustainment, or resupply, requirements to support the LCAAT operation are not trivial. Each LCAAT sortie requires a munitions load of four SDBs, four RATO rockets, a single-use parachute/landing system, and fuel. It also requires LCAAT platform attrition backfills for those lost in combat or to maintenance. These requirements do not vary based on the support options discussed previously, but on

Figure 5.5
Sortie Generation and BOS Equipment Requirement as a Function of Travel Time to Staging Base

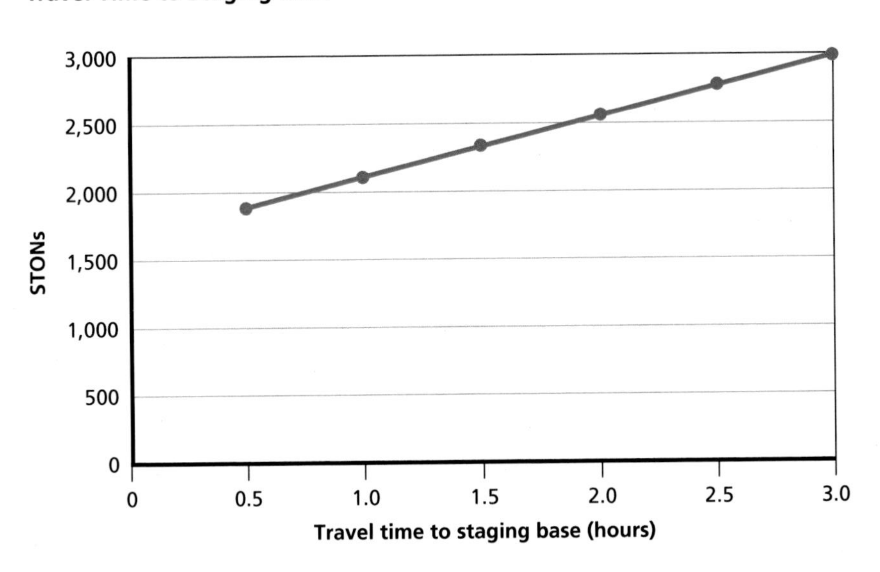

the operational requirements in terms of number of sorties per day, sortie duration, and mission type. Table 5.3 shows the daily sustainment requirements for 864 daily sorties.

Table 5.3
Daily Sustainment Requirements to Support 864 Sorties

Sustainment Type	STONs
RATOs	363
Chute/landing systems	86
Munitions	518 for SDBs
Food/water	23
Fuel	1,378 (411k gal)
LCAAT attrition backfill	108–324

NOTE: Food/water is the daily requirement to support the 1,100 mission generation and CS support personnel as shown in Table 5.1.

The largest contributor to the daily sustainment requirement is fuel. We assumed that the full LCAAT fuel capacity, 476 gallons, was used on every sortie. We assumed a burn rate of 450 pounds per hour, resulting in a sortie duration of approximately seven hours. The sortie duration may vary significantly based on mission needs, and therefore the fuel requirement and associated fuel storage requirements could also vary depending on the mission. The next largest contributor to the daily sustainment requirement is the munitions. Given that we have compared options based on constant combat power provided, in terms of targets prosecuted, the munitions sustainment requirement is necessarily fixed. Finally, RATO rockets and parachute/landing systems required for launch and recovery of the LCAATs are a significant contributor to sustainment requirements. This is an area where design alternatives may exist to reduce the sustainment requirement.

In addition, a daily attrition backfill for the LCAAT platforms would be required. According to inputs from SMEs, the maximum number of sorties expected per LCAAT is ten, which translates to a minimum attrition of 10 percent, or 86 platforms per day. We considered an upper bound of attrition of 30 percent, or 259 platforms per day. This translates to an additional daily sustainment requirement of 108–324 STONs.

Requirements for a 14-Day LCAAT Operation

In the previous sections we calculated the equipment and sustainment requirements under a variety of assumptions on support case, turn time, and launch mechanism. We highlighted that turn time is the primary driver of equipment requirements and that the dependence on RATOs for launch is the primary driver (other than fuel and munitions, which are essentially fixed) for sustainment requirements. To illustrate the full variation in requirements, Figure 5.6 shows the requirement for a 14-day LCAAT operation for three cases: a 12-hour turn time with four RATOs required, a 3-hour turn time with four RATOs required, and a 3-hour turn time with no RATOs required, compared with the F-16. In each case we exclude fuel and munitions weight from the sus-

Figure 5.6
Requirements for 14-Day LCAAT Operation

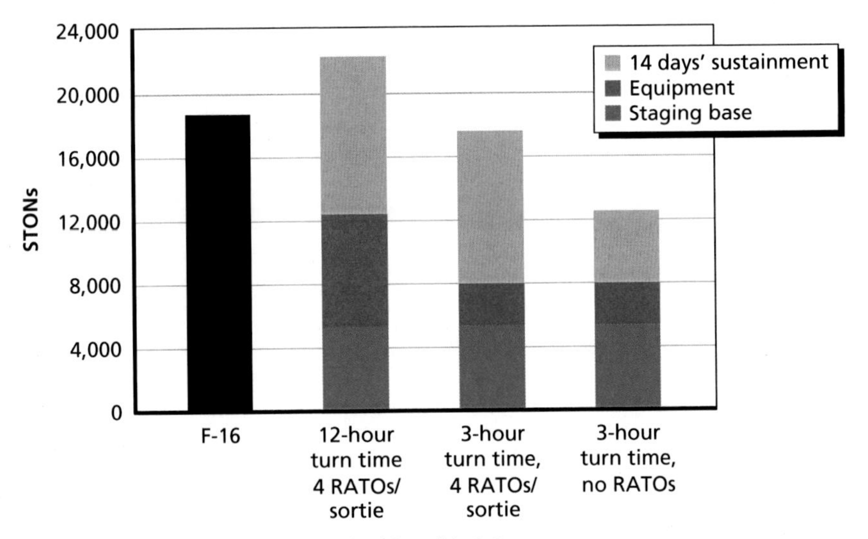

NOTE: F-16 requirements as described in Table 2.2.

tainment requirement. When the sustainment requirements of the LCAAT are included in the total footprint, it compares somewhat less favorably to the F-16. However, even in the worst-case scenario, the footprint is only 20 percent higher, and the F-16 numbers do not incorporate the required tanker and combat search and rescue that would be required, which would be substantial. Additionally, the LCAAT operation is likely more resilient to adversary attack than the conventional F-16 operation. Given the substantial difference in footprint between the best and worst case, design options that reduce turn time and/or reduce or eliminate the need for RATOs should be explored. We highlight a few such options in the next chapter.

CHAPTER SIX

Weapon System Design Considerations

The analysis presented in this report focused primarily on the envisioned capabilities and design of the XQ-58A version of the LCAAT weapon system family and support capabilities currently available. These assumptions achieved the primary objective to operate in an austere environment without relying on a runway or air base for fixed-wing air vehicle operations for combat power projection, although a runway may be required for sustainment support. This requirement and the desire to develop a low-cost, low-risk, near-term functional concept resulted in certain design choices that our analysis has shown greatly affect the footprint and logistical considerations. One of the major design choices centered on the use of RATO for launch and parachutes/airbags for recovery. These methods were chosen as part of the initial concepts identified by the government and the contractor, since they were currently available technology and could be developed and employed quickly and cheaply for this application.

Although RATO launch and parachute recovery are very reasonable ways to approach the operational design of this air vehicle, our analysis has shown that this method for launch and recovery has some downsides for the ability to operate and sustain the air vehicles in the high-intensity warfights envisioned. Referring back to the turn-cycle tasks outlined in Table 3.1, activities relating to RATO rockets and parachutes take significant time relative to other tasks, and turn times are a major driver of the total logistics footprint. In addition, the requirement for four RATO rockets per launch is a major driver for the sustainment footprint, so eliminating the need for rockets would also reduce the logistics footprint. This chapter considers other potential methods to accomplish the launch and recovery portions of the mission in an effort to better determine the tradespace to reduce some of the footprint and logistical issues identified in the analysis above.

In this chapter, we identify other options the USAF may consider for these operations that could reduce some of the downsides associated with the logistical issues presented in earlier chapters of this report. As discussed in the introduction, our focus is on new capabilities and new concepts. Therefore, this chapter looks at new approaches to reduce the logistical burden for operating the XQ-58A and presents some parametric analysis to provide insight into some of the tradespace of designs beyond the XQ-58A.

Launch and Recovery Logistical Issues

As discussed earlier in this report, the concept uses RATO as the low-tech, tried-and-true method for launching the air vehicles. Our analysis has shown that the RATO rockets required for high-tempo operations would account for a large fraction of the logistical requirement for these operations. Further, the trailer launch system and attaching the rockets would be a time-consuming, manpower-intensive, and potentially dangerous operation. Other, potentially more vexing issues exist. Gathering the spent rockets and returning them to the logistics hub provides additional logistical complexity. Spent rockets jettisoned from aircraft shortly after launch could present a fire or other safety issue downrange of the launch sites. Recovery operations relied on a mature concept involving parachutes and airbags. In addition to the RATO launch concept, other concepts are possible. The rest of this chapter examines other options for launch and recovery that may alleviate some of the negative aspects identified as part of the logistical analysis presented earlier.

Launch Options

The major driver for LCAAT launch options is the desire to reduce the dependence of land-based airpower on fixed infrastructure, particularly runways. Several classes of launch techniques have been used for unmanned aerial vehicle (UAV) launches that do not require a runway.[1] These include pneumatic, hydraulic, and bungie-assisted launch, among others. These techniques are useful for launching small, lightweight UAVs that require fairly low launch speeds. Although we suggest further consideration, these may not be viable options for launching the larger, heavier class of air vehicle considered in this analysis. One option, however—the electromagnetic catapult—appears promising and is considered in this section. Another option discussed later in this section may reduce the overall logistical burden of RATO rockets by exploring short-runway operations (e.g., perhaps a stretch of road) and has the potential to reduce the overall RATO rocket requirement. Additionally, hybrid and mixed options should be considered. For example, a lower-tech catapult being used to reduce the number of rockets required for launch could cut the logistical requirement while staying with a potentially lower-tech, lower-maintenance, cheaper option.

Electromagnetic Catapult

The potential use of electromagnetic catapults to launch aircraft has been discussed for many years. Electromagnetic catapults use a linear induction motor to reach the velocity required for aircraft launch. Induction motors have been in wide use for well over a century, and linear induction motors were conceptualized and then developed for transportation applications decades ago. The U.S. Navy (USN) is currently developing/fielding the electromagnetic aircraft launch system (EMALS) for its new class of aircraft carriers, the Gerald R. Ford class. Incorporation of EMALS (vs. steam catapults) is one of the major design changes for this carrier class. Although the first carrier in this class, the Gerald R. Ford (CVN-78), had its share of problems associated with EMALS, the issues appear to have been solved, and EMALS seems to be functioning well.[2] CVN-78 is about halfway through its post-

---

[1] Elia Atkins, Anfibal Ollero, and Antonios Tsourdos, *Unmanned Aircraft Systems*, New York: John Wiley and Sons, 2016.

[2] Issues with the EMALS included power handling, software, and reliability.

delivery test and trials phase, logging nearly 4,000 catapult launches including all types of USN carrier-based aircraft.[3]

One of the most important considerations of catapult launch of an aircraft is the acceleration (g-force) put on the aircraft during launch. The structural limits of the aircraft (and human limits for manned aircraft) restrict the acceptable g-force. From basic Newtonian physics assuming a body is at rest and undergoes a constant acceleration, we know that the square of final velocity is equal to twice the acceleration times the distance through which the body undergoes the acceleration ($v^2 = 2aS$).[4] Therefore, the final speed required by the air vehicle for stable and controllable flight dictates the required trade between acceleration and stroke distance of the catapult. Figure 6.1 shows this relationship graphically for several different acceleration levels.

Figure 6.1
Relationship Between G-Force Required and Stroke Distance

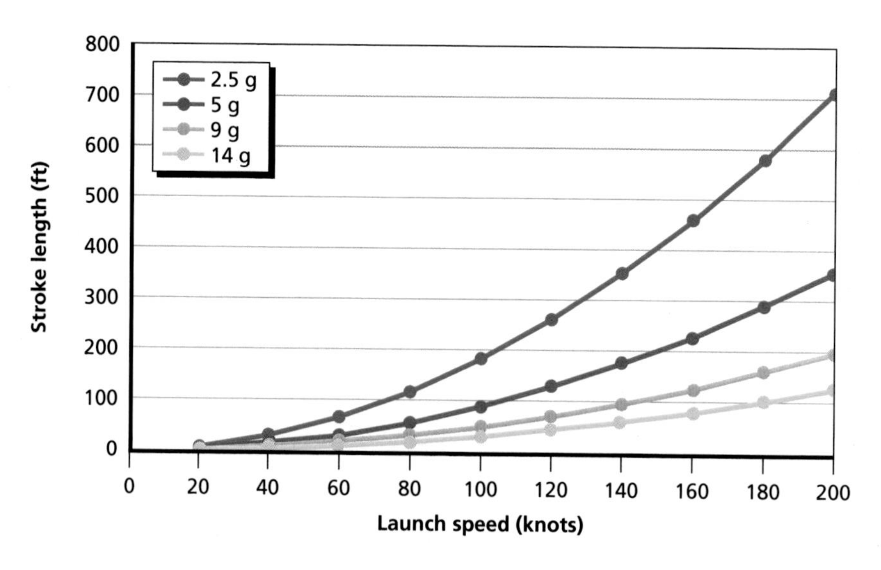

---

[3] United States Navy, "Ford Steams Through Postdelivery Test, Trials," August 7, 2020.

[4] David Holliday and Robert Resnick, *Physics, Parts 1 & 2*, New York: Wiley and Sons, 1978.

This figure shows the nonlinear relationship between stroke length, desired launch velocity, and the required aircraft acceleration. The USN EMALS typically accelerates manned aircraft to an initial speed of 130 to 150 knots, with the g-level limited to around 2.5 to 3 gs for crew and aircraft stress considerations. This requires a stroke length of about 300 ft. A 300-ft stroke-length EMALS is reasonable for carrier operations and is similar in length to the steam catapults that it is replacing. Figure 6.1 provides a great deal of insight into the operational issues surrounding austere LCAAT operations.

The XQ-58A is likely to have a much lower takeoff speed than a typical fighter. In this analysis, we assumed a takeoff speed of 100 knots for the current air vehicle.[5] Assuming the XQ-58A can withstand an acceleration in the 3-to-5-g range, a stroke of 100 to 150 ft would be required to achieve the required 100-knot takeoff speed. Achieving higher accelerations on the LCAAT is not unreasonable. First, launch acceleration is primarily limited for crew considerations. Since LCAAT is unmanned, that constraint is removed. Second, the accelerations on aircraft are limited to reduce the instantaneous structural loads put on the aircraft that can reduce the lifespan and increase the maintenance costs of aircraft. Since the LCAAT is considered an "attritable" aircraft, this constraint can be relaxed as well. EMALS is reportedly capable of achieving 14-g maximum accelerations. Therefore, we will consider developing a launch system for LCAAT that achieves an acceleration of both 9 and 14 gs. Looking at Figure 6.1, this equates stroke lengths from about 30 to 50 ft. This modest required stroke length indicates that electromagnetic launch options for austere operations may be possible. A typical flatbed truck trailer ranges in size up to 53 ft. Therefore, if the air vehicle can withstand accelerations in the 9-to-14-g range, the launch system may fit entirely on a single flatbed trailer or a B-train flatbed (two trailers linked with a fifth wheel to form an 85-ft carrier) to provide a launch system at the lower g-level. One of

---

[5] The XQ-58A with its large aspect ratio to enhance efficiency is closer in design to a Cessna Citation than it is to a high-performance fighter or target drone. The Citation has a stall speed of 42 meters/second. Assuming XQ-58A has a similar stall speed and allowing a 20 percent margin gives about 50 m/s minimum launch speed—approximately 100 knots. We thank our RAND colleague Thomas Hamilton for this insight.

the trade-offs for using a launch system on a flatbed is that it might be less off-road capable; however, in a theater where prepared surfaces are plentiful, launching from a catapult on a flatbed train could be an option to consider.

Developing a modular, flatbed-based, electromagnetic launch system removes many of the problems of employing the system for an austere environment. The system could be thoroughly tested before leaving the main logistics hub. The flatbeds provide the required structural stability for the rail and can also carry the other components (power, energy storage, energy conditioning, cooling, and launch control systems). One could also envision this flatbed system with winches and lifts to load the air vehicle and mount the weapons, providing a full-on completely assembled launch system.

Power Required

A major engineering consideration of an electromagnetic launch system is power generation and especially energy storage. Electromagnetic launch systems impart an enormous amount of kinetic energy (KE) to the air vehicle in a very short period of time. This requires a huge energy storage system that can deliver the required energy over a few seconds. The KE required to launch an air vehicle is[6]

$$\mathrm{KE} = \tfrac{1}{2}\, m\, v^2.$$

Therefore, the required energy to launch a fully loaded 6,000-lb LCAAT (and assuming an additional 100 pounds for a launch carriage system) is about 3.5 megajoules (MJ) per launch to impart an initial speed of 100 knots. Assuming an electromagnetic launch efficiency of 0.9 and a power generation/storage efficiency of 0.5, generating the required power should take about four minutes using a standard 30-kilowatt (kW) generator. So, the power generation appears reasonable in the austere environment given that a standard 30-kW generator could provide the required power between launches. Further, the diesel fuel requirement to run this 30-kW generator (under three gallons per

[6] Holliday and Resnick, 1978.

hour) is very modest.[7] The challenge is likely to be in the energy storage and management systems.

USN EMALS uses a kinetic system for energy storage. In order to launch aircraft in the 100,000-lb takeoff weight class at 150 knots, this flywheel energy storage system would need to be capable of storing about 135 MJ for each launch—roughly 20 times the 6 to 8 MJ required for the operation we are considering in this report. Since the USN system is currently operational, it is likely that many of the engineering challenges have already been overcome, making the application of launching LCAAT much more feasible.

Given that the launch weight for the LCAAT is roughly 5–10 percent of that required to launch a fully loaded carrier-based aircraft and that the launch speed required is about a third less, it may be possible to use a capacitor bank to achieve the required energy storage. Although any design engineering and trade-off analysis is beyond the scope of this report, we do employ some basic physics to explore the potential energy storage requirement here. Initial research indicates that a 2-MJ capacity capacitor bank might be possible using sets of capacitor banks.[8] The 2-MJ design described in this report is roughly one-third of a cubic meter. Therefore, a 4-MJ capacitor bank capable of launching the LCAAT to the required speed may not be out of the question in a package appropriate for this installation and operation from a flatbed trailer.

We present this energy storage discussion to illustrate that multiple options for meeting the requirement to operate in an austere environment may exist. A KE storage system (similar to EMALS) may be preferred for any near-term design. But this is part of the initial engineering design options. In addition to the energy storage system, power conditioning, control systems, cooling and energy production

[7] The 30kW TQG Generator Set, MEP-805A, requires 2.6 gallons per hour. See Generator Set, Diesel Engine, 30kW, 60Hz, TM 12359-OD, 2011.

[8] R. F. Ramazanov, B. E. Fridman, K. S. Kharcheva, O. V. Komarov, and R. A. Serebrov, "Conceptual Design of 2 MJ Capacitive Energy Storage," *Defence Technology*, Vol. 14, No. 5, October 2018.

(discussed above) would be the minimal components required for a self-contained electromagnetic launch system.

The section provided some basic physics and engineering analysis to explore whether an electromagnetic launch system is at all feasible for this application.

Short-Runway Operation

In this section, we consider moving away from the "no runway" assumption to better understand how we might reduce the sustainment burden resulting from the large number of RATO rockets without going to a high-tech solution, such as electromagnetic launch. The trailer-mounted, no-runway launch option requires four 15-KS-1000 Rocket Motor, MK6 Mod 1s.[9] The impulse, defined as the mass times the final velocity, required to launch the air vehicle is about 400,000 Newton-seconds. This assumes a vehicle mass of 6,576 lbs (LCAAT plus four RATO rockets) and a final speed of 100 knots. The impulse generated is defined as a thrust over time—in this case, the LCAAT main engine and four RATO rockets for 15 seconds. The air vehicle engine produces a maximum thrust of about 2,000 pounds of force, and each of the four RATO rockets provides 1,000 pounds of thrust for 15 seconds.[10] Using these data, we compute that about twice the impulse is generated as required. This 40-percent "efficiency" accounts for many factors not included in these simple impulse calculations, such as aerodynamic drag, friction coming off the trailer, misaligned thrust, and others, as well as provides a safety margin to ensure the air vehicle achieves a suitable launch speed. We will use this 40-percent efficiency throughout our calculations to estimate the number of rockets required for short-runway operations.

Our calculations show that even with a 1,000-ft runway, the required impulse (including the 40-percent factor) would still require one or two rockets for launch. These calculations assume that a 200-lb gear trolley is left behind when the rocket(s) fire at the end of the runway but assume no rolling friction from the trolley. It is clear from these

[9] USAF TO 11A11-23-7, 2019.

[10] Air Force Technology, "XQ-58A Valkyrie Unmanned Aerial Vehicle," undated.

simple calculations that even with a fairly long runway, a significant number of rockets will still be required; and depending on the number of 1,000-ft-long runways that may be available in the launch area, it may become a high-value target for the enemy to disrupt operations.

Recovery Options

In this section, we consider ways to eliminate the need for the parachute and airbag recovery system. Parachutes and airbags provide a near-term, low-tech, low-cost solution, but again, these concepts have downsides. As shown earlier in this report, packing and verifying the recovery system requires a significant amount of time and specialized manpower. This process is likely to increase the manpower and air vehicle requirement and/or increase the time required to launch sorties, thereby reducing the overall sortie rate. Further, collection of the parachutes and airbags in the recovery area will need to be addressed. Even if the packing of the parachutes and airbags is done only at the hub, a significant logistical requirement would be imposed, since the recovered items, as well as non-mission-capable recovered air vehicles, would need to be retrograded to the staging base and then brought back to the launch area once repacked or repaired. The parachutes and airbags present a potential logistical burden and will require manpower and possibly equipment to gather, recondition, and install the parachutes and airbags into the air vehicles for subsequent launches. If the parachute could be eliminated, the air vehicle's weight might be reduced, allowing for more fuel or weapons while maintaining the same takeoff weight.

Several options exist for recovering aircraft when a runway is not available. These include some type of net system (used by many light, slow unmanned aerial system [UAS]) or an arresting gear (used by carrier aviation). In addition, using skids like those used on the X-15 program could also be considered, but this would require a length of prepared level space and a dedicated landing area. Retractable skids would also need to be integrated into the LCAAT design, increasing the empty weight and complexity of the design. None of these options seem attractive for this application.

Similarly, using an arresting gear carries a large number of downsides. First, the LCAAT would need landing gear and a tail hook. Incorporating this into the air vehicle would increase the complexity of the design, add additional empty weight that could be significant, especially to distribute the loads required for an arrested landing, and would also increase operational complexity by needing to coordinate all air vehicles and the arresting gear operations rather than simply popping the chute when the air vehicle arrives back at the launch site. Incorporating landing gear into the air vehicle would also involve the complexity of a retraction mechanism and storage within the vehicle to reduce drag, otherwise the range and endurance would be sacrificed. It may also be possible to use a tail hook integrated on the air vehicle and some sort of trolling landing gear to eliminate the need to incorporate landing gear on the aircraft, but this has an enormous amount of complexity and a high risk of damage to the air vehicle when impacting the trolley. Deployment and setup of the arresting gear could also be a challenge under austere conditions.

Net capture systems are often used for small, light UAS that can fly at slow speeds. Since the LCAAT is heavy relative to the typical UAS and has a relatively high stall speed, recovery using a net system appears challenging. The use of nets would have a high chance of damaging the LCAAT and may not even be a viable option given the speeds and weights involved. It may be more plausible to use nets to trap the LCAAT after the parachute has been deployed, thereby eliminating the need for the airbags, but it is hard to envision a net system alone as a recovery mechanism. Depending on the uncertainty and overall relative difficulty of packing parachutes relative to airbags, eliminating the airbags with the use of nets to trap the air vehicle after the parachute is deployed may be a useful option.

Weapon and RATO Handling Options

Another issue we confronted during our analysis was the lack of a desirable weapon loading system. Transporting, handling, and loading the weapons and rockets on the air vehicles is an area where improvements could be made to the initial concept of operation. Although the SDBs and rockets considered in this analysis are capable of being loaded by hand using relatively simple and light equipment (e.g., using a "hernia bar"), this is not optimal. We considered several other methods for munitions loading (e.g., hand jacks), but the primary issue we encountered is that all of this equipment was designed for airfield or prepared surfaces using AM-2 matting.

While it may be possible to find a sufficient number of semi-prepared sites that are large enough for these operations (e.g., soccer fields or other sports fields), which are typically close to roads and which permit the use of standard airfield equipment, this analysis considered the more challenging and flexible case of assuming unprepared locations. The concept envisioned here assumed that the time required to prepare the launch sites to a level that would permit the use of air base weapon handling and loading equipment would not be available. We were unable to identify weapon handling and loading equipment capable of operating in an austere (off-road) environment. In our logistical analysis presented earlier in this report, we relied on the tow vehicles or modified Kubota tractors to transport the expendables from the launch area or cluster storage and to load the weapons and rockets on the air vehicles. We identified some hydraulic lift jacks and other equipment that could help streamline the operation, but typical USAF munitions handlers and loaders designed to operate from advanced air bases and prepared surfaces do not appear to be a viable option for off-road austere operations. Winches (either ones built into the weapon bay or ones that can be attached for loading and then removed prior to flight) and other mechanical devices could help, but these would not streamline operations to any great extent. Conversations with the support equipment and vehicles program office revealed that they were not aware of the requirement for such an off-road, weapon loading capability.

We suggest that the USAF consider better weapon handling equipment for austere and off-road operations. This could include modified, high-mobility, multipurpose wheeled vehicles or other trucks with special handling equipment installed or something akin to an off-road weapon handler and loader—an off-road jammer. We suggest prototyping and experimenting to identify the best way forward in this regard.

Concluding Thoughts

This chapter identified and evaluated potential options to mitigate the downsides and shortfalls identified during our logistical analysis of the current operational concept. Our main takeaways from this work, relative to a 6,000-lb maximum gross takeoff weight (MGTOW) air vehicle (i.e., the XQ-58A), are the following:

- The use of an electromagnetic launch system could be a viable option, owing to recent USN experience solving the technical and engineering challenges, and could significantly reduce the sustainment burden by eliminating the need for rockets.
- Initial calculations on size, volume, and power indicate that a system capable of launching unmanned air vehicles with the characteristics described in this report may fit entirely on one or two flatbed trailers. This would provide an enormous benefit in an austere environment by having a completely self-contained launch system requiring little or no assembly in the field, but may have the limitation of requiring launch from an improved surface if the flatbeds are not off-road capable. We suggest that the USAF do further analysis on this concept to understand the viability for these operations.
- Moving away from a no-runway constraint and allowing short-runway operations could reduce the rocket logistical requirement by one-half to three-quarters, but it has a major downside of providing the enemy with a potential valuable target depending on the number of short stretches of flat pavement available.
- Other hybrid/mixed options should be considered, such as lower-tech catapults (pneumatic, hydraulic, and bungie-assisted), perhaps coupled with fewer rockets than are required for launch operations with no takeoff assistance.
- Although there are a number of options to eliminate the need for parachutes and/or airbags, the recovery flight characteristics of the air vehicle and the austere environment make this challenging.

- Prototyping and experimenting with off-road, austere environment equipment to handle and load expendables (munitions and rockets) on the air vehicles should be pursued and could lead to some efficiency gains in terms of both turn time and subsequently manpower.

CHAPTER SEVEN

Observations and Recommendations

The USAF is exploring the use of a low-cost attritable aircraft weapon system class that could range in cost from $2 million to $20 million. At the same time, it envisions that some instantiations of this class of weapon systems could operate decoupled from fixed infrastructure. For our analysis, we focused on a platform that falls at the lower end of the cost spectrum and is capable of operating independent of a runway. Using the XQ-58A (6,000-lb MGTOW) experimental platform as a basis, our analysis of various logistics and sustainment concepts revealed several observations.[1] They include the following:

- The time required to recover an LCAAT and prepare it for its next mission determines the amount of combat power that an LCAAT-equipped unit can deliver and is the largest determinant of personnel and equipment requirements.
  - Personnel and equipment requirements for LCAAT operations are considerably less than those for traditional platforms.
  - For a similar level of weapon delivery, the LCAAT operation requires approximately 20–60 percent of the personnel and 40–65 percent of the equipment required for a traditional F-16 operation, depending on LCAAT turn time.

[1] It would be worthwhile to conduct similar analysis across the spectrum of sizes and weights for the LCAAT class of weapon system, as other variants could face different limitations and considerations. For example, the challenges associated with RATOs and munitions handling equipment would not be prevalent with a much lighter variant that serves as a sensor platform.

- Sustainment requirements for LCAAT operations are significant. LCAAT takeoff and recovery methods currently being considered by the USAF (i.e., RATO and parachute/airbags) contribute significantly to the sustainment footprint. The trade-off for that footprint is increased resiliency through dispersal and runway independence.
  - Eliminating the need for rockets could reduce the daily sustainment requirements by 50 percent.
- Few traditional USAF capabilities (e.g., BEAR, R-11 refuelers, FORCE) are suited for the type of expeditionary, runway-independent operations envisioned for the LCAAT.
- For a given turn time, alternative, nontraditional capabilities can reduce the footprint and increase resilience by enabling distributed operations (e.g., one BEAR vs. many ARRKs, one FORCE vs. many HERS).
- In considering future force designs, logistics and sustainment analysis early in the process provides benefits to both the research and acquisition communities. It highlights the issues that are barriers to deployment and employment and affords the research community an opportunity to consider engineering design modifications around those areas. For the CS acquisition community, it signals the types of capabilities (e.g., vehicles, munitions loaders) that they should be considering for future operations.

In light of those observations, we offer the following recommendations:

- The USAF should continue to pursue a version of an LCAAT as part of a future force design that is capable of operating decoupled from fixed infrastructure.
- The USAF should continue to explore launch methods that are mobile, do not require a takeoff run, and eliminate the need for RATOs.
- The USAF should also pursue recovery methods that are precision guided and can reduce the time required to prepare the air vehicle for its next mission.

- Given the impact of assembly time and turn time on the manpower required to generate high-volume force projection, the USAF should aggressively pursue ways to shorten the high turn-time drivers. Progress along these lines can best be made by procuring prototype LCAATs and supporting equipment and conducting extensive field tests and experiments.
- The USAF should consider institutionalizing the use of nonmainstream capabilities (e.g., HERS, ARRK) in its tactics, techniques, and procedures for conducting operations absent fixed infrastructure in a manner that improves operational resiliency.
- The USAF should actively engage the CS research (for engineering design considerations) and acquisition (for acquiring uniquely capable equipment) communities in exploring alternative capabilities better suited to supporting the evolving concepts of employment for a future force.
- The USAF should consider institutionalizing the process of analyzing the logistics and sustainment implications of deploying and employing weapon systems being considered in the future force design. Proactively designing for deployment and employment can accelerate the fielding of a viable operational capability.

References

AFPAM—*See* United States Air Force Pamphlet.

Air Force Technology, "XQ-58A Valkyrie Unmanned Aerial Vehicle," undated. As of September 4, 2020: https://www.airforce-technology.com/projects/xq-58a-valkyrie-unmanned-aerial-vehicle/

Atkins, Elia, Anfibal Ollero, and Antonios Tsourdos, *Unmanned Aircraft Systems*, New York: John Wiley and Sons, 2016.

Baron, William, and Douglas Meador, "Low Cost Attritable Aircraft Technology Initiative: XQ-58 Information for CONOPS Analysis," briefing slides, Dayton, Ohio: Air Force Research Laboratory, October 29, 2018. Not available to general public.

Brown, Charles Q., Jr., *Agile Combat Employment (ACE): PACAF Annex to Department of the Air Force Adaptive Operations in Contested Environments*, Honolulu, Hawaii: Headquarters Pacific Air Forces, June 2020.

Clauser, Chris, "Air Rapid Response Kit (ARRK) Capabilities Brief," Air Force Special Operations Command, undated.

Generator Set, Diesel Engine, 30kW, 60Hz, TM 12359-OD, 2011. As of December 16, 2020: https://www.marcorsyscom.marines.mil/Portals/105/pdmeps/docs/MEP/B0953B0971.pdf

Hamilton, Thomas, and David Ochmanek, *Operating Low-Cost, Reusable Unmanned Aerial Vehicles in Contested Environments: Preliminary Evaluation of Operational Concepts*, Santa Monica, Calif.: RAND Corporation, RR-4407-AF, 2020. As of August 18, 2020: https://www.rand.org/pubs/research_reports/RR4407.html

Holliday, David, and Robert Resnick, *Physics, Parts 1 & 2*, New York: Wiley and Sons, 1978.

Krat, Laine F., "Air Force Design/Futures Game 2019: Basing and Logistics Seminar—ACS Concepts," briefing slides, San Antonio, Tex.: Air Force Installation and Mission Support Center, January 11, 2019.

Leftwich, James A., *Low-Cost Attritable Aircraft Technology: Logistics Concept of Support for Deployment and Employment*, Santa Monica, Calif: RAND Corporation, 2020, Not available to the general public.

Mills, Patrick, James A. Leftwich, John G. Drew, Daniel P. Felten, Josh Girardini, John P. Godges, Michael J. Lostumbo, Anu Narayanan, Kristin Van Abel, Jonathan William Welburn, and Anna Jean Wirth, *Building Agile Combat Support Competencies to Enable Evolving Adaptive Basing Concepts*, Santa Monica, Calif.: RAND Corporation, RR-4200-AF, 2020. As of July 20, 2020: https://www.rand.org/pubs/research_reports/RR4200.html

Mills, Patrick, James A. Leftwich, Kristin Van Abel, and Jason Mastbaum, *Estimating Air Force Deployment Requirements for Lean Force Packages: A Methodology and Decision Support Tool Prototype*, Santa Monica, Calif.: RAND Corporation, RR-1855-AF, 2017. As of January 6, 2020: https://www.rand.org/pubs/research_reports/RR1855.html

Ramazanov, R. F., B. E. Fridman, K. S. Kharcheva, O. V. Komarov, and R. A. Serebrov, "Conceptual Design of 2 MJ Capacitive Energy Storage," *Defence Technology*, Vol. 14, No. 5, October 2018, pp. 622–627. As of September 4, 2020: https://www.sciencedirect.com/science/article/pii/S2214914718303696

Rampmaster Corporation, "17,500 Gallon WD Modular Lift Deck Jet Refueler," undated. As of December 15, 2020: https://www.rampmasters.com/refueler-solutions/17500-gallon/

Trimble, Steve, "USAF Defines Price Range for 'Attritable' UAS," *Aviation Week*, August 3, 2020. As of September 4, 2020: https://aviationweek.com/defense-space/aircraft-propulsion/usaf-defines-price-range-attritable-uas

United States Air Force Handbook 10-222, Volume 2, *Bare Base Assets*, February 6, 2012. As of April 22, 2021: https://www.wbdg.org/FFC/AF/AFH/afh10_222_v2.pdf

United States Air Force Pamphlet 23-221, *Materiel Management—Fuels Logistics Planning*, March 11, 2013. As of April 22, 2021: https://static.e-publishing.af.mil/production/1/af_a4_7/publication/afpam23-221/afpam23-221.pdf

United States Air Force Tactics, Techniques, and Procedures 3-34.1, *Services Contingency Beddown and Sustainment*, November 1, 2007. As of April 22, 2021: https://static.e-publishing.af.mil/production/1/af_a1/publication/afttp3-34.1/afttp3-34.1.pdf

United States Department of Defense, *DoD Dictionary of Military and Associated Terms*, June 2020. As of December 9, 2020:
https://www.jcs.mil/Portals/36/Documents/Doctrine/pubs/dictionary.pdf

United States Navy, "Ford Steams Through Postdelivery Test, Trials," August 7, 2020. As of May 3, 2021:
https://www.navy.mil/Press-Office/News-Stories/Article/2303562/ford-steams-through-postdelivery-test-trials/

USAF TO—*See* United States Air Force Technical Order.

Varden, Richard, "PACAF Civil Engineer Force Modeling Adaptive Basing: Minimum Endurance," PowerPoint presentation, Joint Base Pearl Harbor-Hickam, Honolulu, Hawai'i: Headquarters Pacific Air Forces, October 4, 2016, Not available to the general public.

Vick, Alan J., *Air Base Attacks and Defensive Counters: Historical Lessons and Future Challenges*, Santa Monica, Calif.: RAND Corporation, RR-968-AF, 2015. As of September 29, 2020:
https://www.rand.org/pubs/research_reports/RR968.html